First International Conference on Virtual Education (ViEduc2018)

Challenges and Opportunities

Journal of Physics: Conference Series
Volume 1161

Bucaramanga, Colombia
8 – 12 October 2018

ISBN: 978-1-5108-8210-2
ISSN: 1742-6588

Printed from e-media with permission by:

Curran Associates, Inc.
57 Morehouse Lane
Red Hook, NY 12571

Some format issues inherent in the e-media version may also appear in this print version.

Printed by Curran Associates, Inc. (2019)

For permission requests, please contact the Institute of Physics at the address below.

Institute of Physics
Dirac House, Temple Back
Bristol BS1 6BE UK

Phone: 44 1 17 929 7481
Fax: 44 1 17 920 0979

techtracking@iop.org

Additional copies of this publication are available from:

Curran Associates, Inc.
57 Morehouse Lane
Red Hook, NY 12571 USA
Phone: 845-758-0400
Fax: 845-758-2633
Email: curran@proceedings.com
Web: www.proceedings.com

Table of contents

Volume 1161

First International Conference on Virtual Education: Challenges and Opportunities

8–12 October 2018, Bucaramanga, Colombia

Accepted papers received: 18 December 2018
Published online: 26 February 2019

Preface

Papers

Teaching, Learning and Evaluation Strategies Mediated with TIC

Academic and Administrative Management of Virtual Education

Pedagogical Models in Virtual Education

Virtual Education Platforms

Digital Educational Resources

IOP Conf. Series: Journal of Physics: Conf. Series **1161** (2019) 011001 doi:10.1088/1742-6596/1161/1/011001

Preface: First International Conference on Virtual Education. Challenges and Opportunities

D A Miranda[1] **and E D V-Niño**[2,3]
[1] Universidad Industrial de Santander, Bucaramanga, Colombia
[2] Universidad Politécnica de Madrid, Madrid, Spain
[3] Foundation of Researchers in Science and Technology of Materials, Colombia

E-mail: deydannv@gmail.com, dalemir@uis.edu.co

The first International Congress on Virtual Education, Challenges and Opportunities, ViEduc2018, was doing in Bucaramanga between 8 and 12th October of 2018, as part of the commemoration of the 70th anniversary of the "Universidad Industrial de Santander", UIS, with the participation of 230 participants. This Congress was articulated with different events around the theme of Education and ICT to discuss the positive impacts of the use and appropriation of technologies in education, the transformations they generate, their possibilities and challenges. The Institute for Regional Projection and Distance Education, iPred, the Academic Vice-Chancellorship and the School of Education and the Center for the Development of Teaching, CEDEDUIS, at the "Universidad Industrial de Santander," UIS, were the organizer of this Congress.

ViEduc2018 was a space for academic and scientific exchange focused into five topics: teaching and learning strategies supported by ICT; academic and administrative management of virtual education; pedagogical models in virtual education; virtual education platforms and digital education resources. These topics were aborted by 11 Magistral Conferences presented by professors of Indiana University Purdue University Indianapolis (USA), "Universidad de Sevilla" (Spain), University of Cranfield (United Kingdom), Open University (United Kingdom), Thapar Institute of Engineering & Technology (India), "Universidad del Externado" (Colombia), "Universidad de Antioquia" (Colombia), "Consejo Nacional de Acreditación, CNA" (Colombia), "Universidad Nacional Abierta y a Distancia, UNAD" (Colombia), "Politécnico Grancolombiano" (Colombia) and "Universidad UNIMINUTO" (Colombia). Besides, 22 research works were presented as Oral presentation and 21, as Posters and one forum about the challenges in and opportunities about virtual and distance education.

The papers contained in this edition are original contributions to the topics described before and shown the academical works that is doing in the field of virtual and distance education.

IOP Conf. Series: Journal of Physics: Conf. Series **1161** (2019) 011001 doi:10.1088/1742-6596/1161/1/011001

Organizing Committee
David Alejandro Miranda Mercado (Chairman)
Claudia Patricia Calderón Lozano

International Committee

- Gregor Novak, Indiana University-Purdue, Indianapolis, and United States Air Force Academy, USA
- Konstantinos Salonitis, Cranfield University, United Kingdom
- Satheesh Krishnamurthy, The Open University, United Kingdom
- Susheel Mittal, Thapar University of Engineering & Technology, India
- Ph.D. Ely Dannier Valbuena Niño, Laboratorio de Desarrollo de Productos, Universidad Politécnica de Madrid, Spain

National Committees

- Adriana Rocío Lizcano Dallo, Universidad Industrial de Santander, Colombia
- Jorge Winston Barbosa Chacón, Universidad Industrial de Santander, Colombia
- David Alejandro Miranda Mercado, Universidad Industrial de Santander, Colombia

IOP Conf. Series: Journal of Physics: Conf. Series **1161** (2019) 011001 doi:10.1088/1742-6596/1161/1/011001

4:20 p. m. - 4:40 p. m. Ágora. Edificio Facultad de Ciencias Humanas
GAEV-03: La educación virtual desde el enfoque de las capacidades humanas.
M. García

4:40 p. m. - 5:00 p. m.
GAEV-05: Articulación de las sedes regionales de la Universidad Industrial de Santander para un ciclo básico de ingeniería unificado.
D. Miranda, G. Patiño, A. Uzcano, R. Ospina, J. Martínez, M.Sánchez, E. Castellanos, R. Valdivieso

5:00 p. m. - 5:20 p. m.
MPEV-03: La formación online como estrategia de la escuela de ciencias administrativas y económicas de UniMonserrate: un espacio para conocer y aprender.
I. Osma, C. Martínez

5:20 p. m. - 5:40 p. m.
RED-09: Diseño e implementación de una estrategia para el acompañamiento y fortalecimiento de competencias blandas para emprendedores a partir del uso del App Estoy Bien.
L. Villamizar, L. Vera, W. Villamil, O. Yabur

5:40 p. m. - 6:00 p. m.
MPEV-11: Actividades académicas estratégicas para el fortalecimiento de competencias y habilidades de los estudiantes de la Unad: caso fábrica de café y chocolate El Agrario, de San Vicente de Chucurí, Santander, Colombia.
I. Medina, L. Quintana, A. García

Viernes 12 de octubre de 2018

8:00 a. m. - 9:00 a. m. Ágora. Edificio Facultad de Ciencias Humanas
Conferencia magistral: Criterios para evaluar la calidad de programas virtuales y a distancia. Un proceso de construcción colectiva como insumo para la definición de lineamientos de acreditación en alta calidad en Colombia.
Profesor Fernando Cantor, CNA

9:00 a. m. - 10:00 a. m.
Conferencia magistral: Desafíos de la educación virtual y a distancia en la sociedad del conocimiento: claves para entender su impacto social educativo.
Profesor Jaime Alberto Leal Afanador, Universidad Nacional Abierta y a Distancia.

10:00 a. m. -10:30 a. m. Plazoleta entre Ingeniería Mecánica y Camilo Torres
Refrigerio
Seccion de pósteres

10:30 a. m. - 11:30 a. m. Ágora. Edificio Facultad de Ciencias Humanas
Conferencia magistral: Uniminuto virtual y a distancia. Un reto hecho realidad.
Profesora Marelen Castillo Torres, Universidad Uniminuto

11:30 a. m. - 12:30 p. m. Ágora. Edificio Facultad de Ciencias Humanas
Foro Retos y oportunidades de la educación virtual

12:30 p. m. -1:00 p. m. Ágora. Edificio Facultad de Ciencias Humanas
Acto de clausura Primer Congreso Internacional de Educación Virtual: retos y oportunidades

IOP Conf. Series: Journal of Physics: Conf. Series **1161** (2019) 011001 doi:10.1088/1742-6596/1161/1/011001

Lunes 8 de octubre de 2018

8:00 a. m.- 8:30 a. m. — Ágora. Edificio Facultad de Ciencias Humanas

Acto inaugural Semana TIC en Educación.
Alta dirección de la Universidad

8:30 a. m. - 12:00 m. — Audtorio Centic4 piso

Taller 1: MiniMOOC's.
Profesor Susheel Mittal, Thapar Institute of engineering and technology by CTD, Thapar University

12: 00 a. m. - 2:00 p. m. — Almuerzo libre

2:00 p. m. - 6:00 p. m. — Sala Centic 202-203

Taller 2: Preparation of students for interactive lessons.
Profesor Gregor Novak, United States Air Force Academy

Martes 9 de octubre de 2018

8:00 a. m. - 12:00 m. — Sala Centic 302-303

Taller 3: Online manufacturing post-graduate teaching methods and challenges.
Profesor Konstantinos Salonitis, Cranfield University

12:00 a. m.- 2:00 p .m. — Almuerzo libre

2:00 p. m. - 6:00 p. m. — Sala Centic 302-303

Taller 4: Renewable energy teaching methods.
Profesor Satheesh Krishnamurthy, The Open University

Miércoles 10 de octubre de 2018

8:00 a. m. -8:45 a. m. — Ágora. Edificio Facultad de Ciencias Humanas

Conferencia magistral: Flipping instruction with just-in-time teaching.
Profesor Gregor Novak, United States Air Force Academy

8:45 a. m. - 9:30 a. m.

Conferencia magistral: El *e-learning*, alternativa para fortalecer el desarrollo cultural regional: el caso de la especialización en sostenibilidad de museos de la Universidad Externado de Colombia.
Profesora Claudia Villafañe Casadiego, Universidad Externado

9:30 a. m. -10:10 a. m. — Plazoleta entre Ingeniería Mecánica y Camilo Torres

Refrigerio
Seccion de pósteres

10:10 a. m.-10:30 a. m. — Ágora. Edificio Facultad de Ciencias Humanas

GAEV-01: Educación inclusiva y virtualidad: aplicación del índice de inclusión para la educación superior a un programa virtual.
A. Modragon

10:30 a. m.- 10:50 a. m.

MPEV-07: La competencia digital oculta en el falso nativo digital en el nivel superior.
J. Martínez, G. López, I. Guevara, V. Rodríguez, A. Landa

10:50 a. m. - 11:10 a. m.

EEAE-03: Herramientas de aprendizaje basadas en juegos digitales como estrategia para la enseñanza de estadística en estudiantes de Ingeniería Industrial.
H. Romero, E. Fajardo, E. Sánchez, L. Beleño

11:10 a. m. - 11:30 a. m.

RED-04: El aula invertida y su aplicación en la asignatura de análisis financiero: evidencia empírica para la mejora de la cultura financiera.
D. Prada, A. Acevedo, J. Montilla, W. Figueroa, A. Diaz, G. Ardila

11:30 a. m. - 11:50 a. m.

EEAE-05: La importancia de la conceptualización en el uso de Wolfram Mathematica 10 para un curso de cálculo diferencial.
S. Montoya, D. Prada

11:50 a. m. - 12:10 p. m.

EEAE-06: Student perception on the implementation of a teaching strategy based on just in time teaching (JITT), mediated learning and the use of information and communication technologies (ICT) in the Physics I Laboratory course.
F. Pinzón, A. Lizcano, J. Martínez, G. Patiño, D. Miranda

12:10 p. m.- 2:00 p. m. — Almuerzo libre

2:00 p. m. - 2:45 p. m. — Ágora. Edificio Facultad de Ciencias Humanas

Conferencia magistral: Industry and university collaboration for post-graduate manufacturing education.
Profesor Konstantinos Salonitis, Cranfield University

2:45 p. m. - 3:30 p. m.

Conferencia magistral: Experiencia de integración de tecnologías en la Universidad de Antioquia.
Profesora Donna del Socorro Zapata Zapata, Universidad de Antioquia

3:30 p. m. -4:20 p. m. — Plazoleta entre Ingeniería Mecánica y Camilo Torres

Refrigerio
Seccion de pósteres

4:20 p. m. - 4:40 p. m. — Ágora. Edificio Facultad de Ciencias Humanas

MPEV-01: Virtual tool for training and evaluation of problem-based engineering subjects.
A. González, I. Cáceres, W. Pinto

4:40 p. m. -5:00 p. m.

EEAE-08: Análisis de la estrategia de aprendizaje basado en proyectos (ABP) que implementan los docentes en la formación de estudiantes en el curso disciplinar de Gestión de Calidad del programa de Tecnología en Regencia de Farmacia de la Universidad Industrial de Santander.
L. Rincón, N. Rico

5:00 p. m. - 5:20 p. m.

RED-08: TelEduc: Un ambiente virtual para la enseñanza y el aprendizaje en la Universidad de Campinas.
J. Acevedo-Rincón, C. E. Flórez

5:20 p. m. - 5:40 p. m.

PEV-2: Entrenamiento en línea para el fortalecimiento de presaberes matemáticos mediado por la plataforma web Khan-Academy.
K. Rueda, O. Serrano

5:40 p. m. - 6:00 p. m.

MPEV-08: Estrategias pedagógicas de procesos en salud, la reanimación cardiopulmonar básica mediante un ambiente virtual de aprendizaje.
M. Chávez, J. Primo, A. Vara

Jueves 11 de octubre de 2018

8:00 a. m. - 8:45 a. m. — Ágora. Edificio Facultad de Ciencias Humanas

Conferencia magistral: Mini-MOOC's, a concept of precise learning. Profesor Susheel Mittal, Thapar Institute of engineering and technology
by CTQ, Thapar University

8:45 a. m. - 9:30 a. m.

Conferencia magistral: Factores de éxito asociados a la metodología de la educación virtual.
Profesor Julio Cabero, Universidad de Sevilla España

9:30 a. m.10:00 a. m. — Plazoleta entre Ingeniería Mecánica y Camilo Torres

Refrigerio
Seccion de pósteres

10:00 a. m. - 10:20 a. m. — Ágora. Edificio Facultad de Ciencias Humanas

EEAE -13: Implementation of JITT methodology in mathematics and natural sciences study in high school education institutions of Colombian east.
E. Castellanos, D. Miranda, J. Martínez, R. Valdivieso, E. Vera

10:20 a. m. - 10:40 a. m.

RED-11: A model of didactic guidelines to train rural inhabits in the implementation of biodigesters.
J. Quiroga, K. Tschierch, O. Bohórquez

10:40 a. m. - 11:00 a. m.

RED- 01:Google classroom como herramienta mediadora del aprendizaje.
K. Cristiano, D, Triana

11:00 a. m. - 11:20 a. m.

RED- 02: La teleclase como herramienta en la formación virtual de tecnólogos en regencia de farmacia Unab-CES.
C. Martínez, I. Osma

11:20 a. m. - 11:40 a. m.

EEAE -01: Development of a virtual learning environment for the subject numerical methods under Moodle.
A. Becerra, M. Díaz, O. González

11:40 a. m. - 12:00 m.

RED-07: Desarrollos en realidad aumentada para crear experiencias significativas.
G. Figueroa

12:00 m. - 2:00 p. m. — Almuerzo Libre

2:00 p. m. - 2:45 p. m. — Ágora. Edificio Facultad de Ciencias Humanas

Conferencia magistral: Nanotechnology exploring endless possibilities in education.
Profesor Satheesh Krishnamurthy, The Open University

2:45 p. m. -3:30 p. m.

Conferencia magistral: El bienestar universitario en la virtualidad: caso del Politécnico Grancolombiano.
Profesora María del Socorro Guzmán Serna, Politécnico Grancolombiano

3:30 p. m.4:20 p. m. — Plazoleta entre Ingeniería Mecánica y Camilo Torres

Refrigerio
Seccion de pósteres

IOP Conf. Series: Journal of Physics: Conf. Series **1161** (2019) 011001 doi:10.1088/1742-6596/1161/1/011001

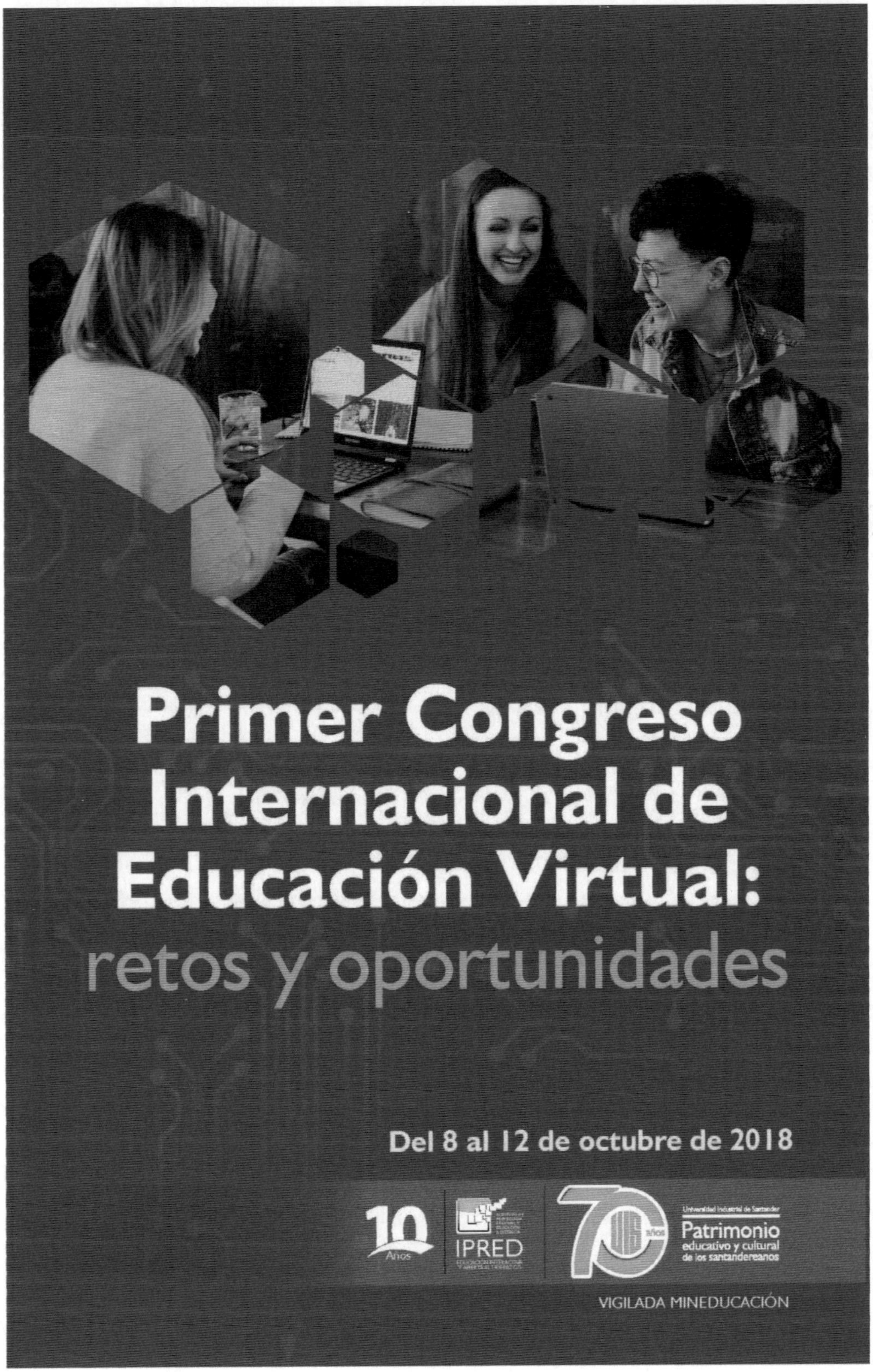

Peer review statement

All papers published in this volume of *Journal of Physics: Conference Series* have been peer reviewed through processes administered by the proceedings Editors. Reviews were conducted by expert referees to the professional and scientific standards expected of a proceedings journal published by IOP Publishing.

IOP Conf. Series: Journal of Physics: Conf. Series **1161** (2019) 012001 doi:10.1088/1742-6596/1161/1/012001

Construction of public policy of higher education with participation of the academic community, Colombian, case for recognition of virtual and distance learning programs

F Cantor[1]
[1] Consejo Nacional de Acreditación, Bogota, Colombia

E-mail: cantorrf@icloud.com

Abstract. As of this date, 8641 academic programs are authorized by Colombia's National Ministry of Education. Of these, 7923 are in-person, 404 are virtual, and 314 are offered at distance. Of the total number of virtual and at-distance programs, only 26 (3.62%) are accredited in the high-quality rank. This low accreditation figure is attributed to a lack of evaluation instruments encompassing criteria to recognize high quality of virtual and at-distance programs. Given the previous, the academic community joined efforts to design and develop the program "Collective construction of criteria aimed at enabling recognition of high quality virtual and at-distance learning programs in Colombia". Regional workshops were conducted that succeeded in gathering representatives of around 100 Higher Education Institutions representing nearly 95% of institutions offering virtual and at-distance learning programs in Colombia. The product of those workshops was consolidated and brought to the consideration of the Principals of those institutions for their respective validation. The validated version was submitted to the National Ministry of Education as a technical document that will support the public policy that rules accreditation of virtual and at-distance learning programs in Colombia.

1. Introduction

Online offer of high education (virtual, at-distance, and mixed) is a reality strongly on the rise at the world level [1, 2]. It suits people whose time, duties or place constraints prevent them from accessing higher education. In the case of higher education institutions, it stands as an alternative to increase their academic offer nationally and internationally. For governments of countries with low schooling, it represents an alternative to enable high education coverage of regions away from the large urban centres that concentrate the main higher education institutions [3, 4]. Colombia has 8641 learning programs authorized by the National Ministry of Education, as shown in Table 1. According to the previous, of the total number of learning programs authorized by the MOE in Colombia, 718 are offered through the Virtual and At-distance (V&D) modalities, and only 26 of them are accredited (3.62%). The previous means that 96% of V&D programs offered are not accredited as high quality. In order to qualify for the evaluation process towards accreditation, learning programs must have been used for a minimum time, a requirement not met by one part of this offer segment, but the decision to seek high quality accreditation is volunteer, without any obligatoriness [5].

Consequently, of the total V&D programs in Colombia, some can meet the life-time requirement and quality conditions but have failed to start accreditation process. Also, there are cases where quality and life-time conditions for evaluation process are met, and the IES unilaterally starts process, without reaching expected results, a situation potentially arising from two situations: 1) The program definitely

IOP Conf. Series: Journal of Physics: Conf. Series **1161** (2019) 012001 doi:10.1088/1742-6596/1161/1/012001

fails to meet quality conditions and must improve the areas highlighted by the assessment team, and 2) the assessment team failed to determine that the program does meet the conditions.

Table 1. Programs authorized by the National Ministry of Education. QR (Qualified Register), administrative act granted by the National Ministry of Education, authorizing offer of an Academic Program. HQA (High Quality Accreditation), administrative act granted by the National Ministry of Education certifying before society that the Program meets quality conditions.

Formation level		In-person	Virtual	At-distance	Total
Technical professional institution	QR	361	29	11	401
	HQA	2	0	0	2
	Accredited percentages	0.55	0.00	0.00	0.50
Technological institution	QR	558	47	27	632
	HQC	20	0	1	21
	Accredited percentages	3.58	0.00	3.70	3.32
University-level institution	QR	1640	160	82	1882
	HQA	161	3	3	167
	Accredited percentages	9.81	1.88	3.65	8.87
University	QR	5364	168	194	5726
	HQA	1125	2	17	1144
	Accredited percentages	20.97	1.19	8.76	20
Total offer	QR	7923	404	314	8641
	HQA	1308	5	21	1334
	Accredited percentages	16.5	1.23	6.68	15.43

Upon debate of this situation, the national academic community concluded that despite the likely occurrence of situations associated with hypothesis 1, the low accreditation number can result from failure of evaluating team to identify compliance with quality conditions. Another conclusion was that V&D learning programs in Colombia are being evaluated through instruments built for accreditation of in-person modality programs. Despite the fact that no difference should exist between professional education objectives, and also that the programs of the different modalities (in-person, virtual, at-distance) must be evaluated through the same quality criteria, it was determined that evaluation instruments aiming at accreditation for programs offered as V&D must contemplate parameters that recognize the specificities of such modalities. In the absence of such specific criteria, evaluators are unable to recognize those specificities and end up applying evaluation criteria they apply for in-person programs, to the detriment of the quality of V&D programs. This situation led to the need to start a collective construction process with the participation of the national academic community that is part of the academic processes of V&D programs in the different higher education institutions.

2. Methodology

The project was carried out by the National Accreditation Council in the following stages:

2.1. Identification and participation of a group

Of a total of 297 academic institutions, around 100 represent the bulk of V&D programs. Along the second semester of 2017, academic representatives of these institutions were invited to join efforts with the Consejo Nacional de acreditación to diagnose the situation of V&D academic offer in the country.

2.2. Hold an international seminar that permits hands-on participation of the national academic community in elaboration of good practices in self-evaluation and evaluation of V&D programs
The representatives of Stage 1 participated in planning and execution of the project "Collective construction of criteria that will enable recognition of V&D learning programs in Colombia".

2.3. Development of regional workshops for establishment at the national level of specific criteria to evaluate the quality of V&D learning programs in Colombia
Along 2018, regional workshops were conducted throughout Colombia, with participation of representatives of 89 higher education institutions (HIE), as shown in Figure 1.

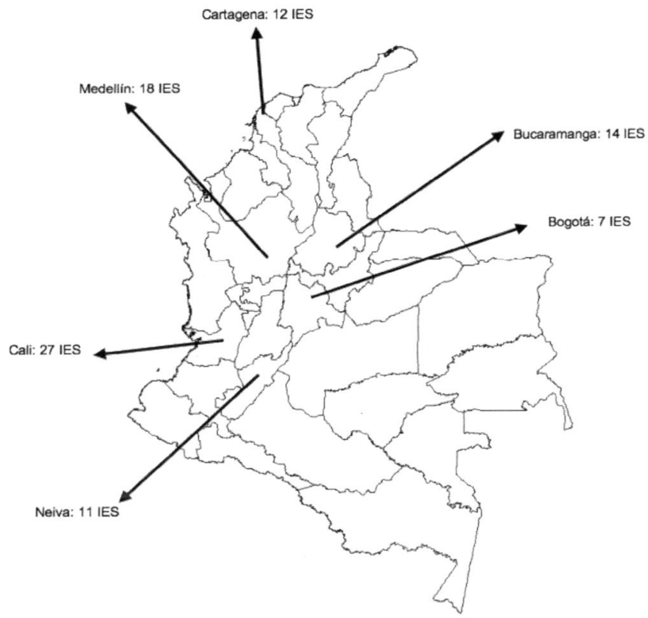

Figure 1. Geographic distribution of the regional workshops conducted in Colombian territory.

2.4. Construction of a unified evaluation criteria proposal of V&D programs incorporating the regional contributions
Deriving from the activity of each regional workshop, different proposals were generated and unified by the Coordinating team, incorporating not only commonly expressed criteria, but also criteria seldom repeated that contributed to characterization of V&D program specificities for Colombia's academic community.

2.5. Submitting the unified proposal to the analysis of the academic community
The unified version corresponding to the collective construction of criteria to recognize the specificity of V&D programs was submitted to the principals of the different academic entities participating in the process.

2.6. Presentation to the National Ministry of Education of a unified proposal that will allow evaluation of the quality of V&D Learning programs in Colombia, with a contribution from the national academic community
The unified version, corrected through contributions of the expert teams proposed by the school principals, was presented to the Ministry of National Education as a collective construction work of the national academic community.

IOP Conf. Series: Journal of Physics: Conf. Series **1161** (2019) 012001 doi:10.1088/1742-6596/1161/1/012001

3. Results and discussion

3.1. Hold an international seminar that permits hands-on participation of the national academic community in elaboration of good practices in self-evaluation and evaluation of V&D programs

The project started with an international seminar carried out in the first semester of 2018, attended by national and foreign experts who discussed the challenges facing the process. Around 70 of the 110 higher education institutions offering V&D learning programs in Colombia attended. The project route map was presented, which called for participation in different regional workshops, and announced the adoption in the regional workshops of the methodology of "rubrics" or "matrixes of evaluation through referents" proposed by the National Ministry of Education in 2018 [5], which permits identification of the descriptors or criteria of evaluation of educational programs and institutions at different stages of consolidation or maturation.

Table 2. Quality conditions considered in the regional workshops.

Quality conditions	Number of features associated with each quality condition	Qualified Register - QR (First time)	Renewal of Qualified Register QRR	High quality accreditation - HQA (First time)	Renovation of High Quality Accreditation – RHQA
At institutional level					
1. Students	4				
2. Teachers	2				
3. Graduates	2				
4. Research	3				
5. Well-being	3				
6. Institutional government	5				
7. Quality planning and improvement	2				
8. Administrative management	6				
9. Infrastructure	2				
10. Financial resources	3				
At Program level					
1. Teachers	2				
2. Curricular management	6				
3. Educational media	2				
4. Extension	2				
5. National and international Interaction	2				

3.2. Development of regional workshops for establishment at the national level of specific criteria to evaluate the quality of V&D learning programs in Colombia

Participants in the regional workshops were asked to describe in each one of the characteristics associated with quality conditions a) at institutional level, and b) at the Program level, the specific criteria offered under the modalities of V&D, taking into account the Program's different maturation stages, from qualified register to (QR) to High Quality Accreditation Renovation (HQAR). Accordingly, the matrixes developed in the different workshops took into account the quality conditions shown in Table 2.

3.3. Submitting the unified proposal to the analysis of the academic community

The unified version corresponding to the collective construction of criteria to recognize the specificity of V&D programs was submitted to the principals of the different academic entities participating in the process for revision of the whole process by the expert teams of each institution that received the document. Prudential lapse was provided to receive observations to the document, after which it was reviewed by the regional coordinators, who produced a second unified collective proposal.

3.4. Presentation to the National Ministry of Education of a unified proposal that will allow evaluation of the quality of V&D Learning programs in Colombia, with a contribution from the national academic community

The unified version, corrected through contributions of the expert teams proposed by the school principals, was presented to the Ministry of National Education as a collective construction work of the national academic community to identify the specific criteria that will allow evaluation of the quality of V&D Learning programs in Colombia, from the stage of Qualified Register, first time, to Renovation of High Quality Accreditation. At present, the Minister of National Education considers this document to be an important tool to generate the next Public Policy instruments that have been approved through administrative act, to guide the self-evaluation and evaluation processes of the quality of V&D Learning programs in Colombia.

4. Conclusions

This work ratifies the fact that a public policy process is amply recognized and welcomed inasmuch as this policy not only reflects the feeling of the recipient community, but also shows the logics of the community's dynamics. In the case of this program, those criteria that allow recognition of the high quality of V&D Learning programs in Colombia are widely welcomed and supported by the academic community whose contributions make up the proposal and set a pathway for continued improvement corresponding to the logic and dynamics of processes inherent to virtual and at-distance learning. This work is the result of a process developed by the National Accreditation Council of Colombia, in a collective process of the National Accreditation Council of Colombia, in a collective construction effort by the national academic community, represented by participants from 112 higher education institutions and organizations, through regional workshops.

Acknowledgements

The author expresses sincere thanks to the national academic community, and mainly to the following professionals who coordinated the regional workshops led by the National Accreditation Council: Arango Sandra - Universidad de Medellín, Figueroa Gloria - Universidad Pontificia Bolivariana, Gallego Germán – Universidad Autónoma de Occidente, Guzmán María del Socorro – Institución Universitaria Politécnico Grancolombiano, Miranda David – Universidad Industrial de Santander, Narváez Lesly – Asociación Colombiana de Instituciones de Educación Superior con Programas a Distancia y Universidad Nacional Abierta y a Distancia, Paque Ana Lucía – Corporación Universitaria del Huila, y Vergara Clara – Universidad de Cartagena.

References

[1] Martínez Castro R 2018 *Investicga* **2** 81-89
[2] Klement M 2017 *Computers & Education* **105** 31-43
[3] Pandey S and Tiwari S 2014 *Procedia - Social and Behavioral Sciences* **157** 55-62
[4] Mora H, Pujol Lopez F A, Mendoza Tello J C and Morales Morales M R 2018 *Computers in Human Behaviour* **In Press**
[5] Ministerio de Educación Nacional 2018. *Referentes de Calidad: una propuesta para la evolución del Sistema de Aseguramiento de la Calidad* (Bogota: Imprenta Nacional)

IOP Conf. Series: Journal of Physics: Conf. Series **1161** (2019) 012002 doi:10.1088/1742-6596/1161/1/012002

Active learning pedagogy: Structuring the pre-instruction assignment

G Novak[1,2,3]
[1] Indiana University, Indianapolis, USA
[2] Purdue University, Indianapolis, USA
[3] United States Air Force Academy, Colorado, USA

E-mail: gnovak@iupui.edu

Abstract. Education research is changing the way teachers teach and students learn. In contrast with the lecture-based approach to instruction, centered on the teacher as "performer", teaching and learning activities now place equal emphasis on in-class and out-of-class work. Much of the interaction between at-home activity and classroom experience and is mediated by information technology tools. Students are technology savvy, comfortable in the world of on-line delivery of information and social media interactions. To be an effective player in this world, instructional design has to follow the students. Contrary to the opinion that the introduction of technology into the teaching and learning process will diminish the role of the teacher, we believe that the opposite could be the case. Communication technology can enable all students to be exposed to the best and most charismatic lecturers. Electronically enabled social media forums can give rise to much more individualized interactions between students and instructors and students and students than what was ever possible. In this paper, we focus on some pedagogical approaches that allow the students to prepare themselves for interactive classroom engagement. The techniques we describe are labelled Just-in-Time Teaching and Worked-Examples. These involve carefully crafted assignments that students complete before going to class. The results of these assignments, completed just before class-time and available to the instructor, prepare the both for a more productive in-class experience.

1. Introduction

Over the past several decades teaching and learning activities have shifted from passive, instructor-centered to active learner-centered approaches [1] Variously labelled active learning or interactive engagement, the main point of the strategy is "that it involves students in doing things and thinking about the things they are doing. [2]." Today's classrooms include a kaleidoscope of "non-traditional" students: older students, working part-time students, commuting students, and, at the service academies, military cadets. Providing consistent, friendly support can make the difference between a successful experience and a fruitless effort. It can even mean the difference between graduating and dropping out. Education research has made us more aware of learning style differences and of the importance of passing some control of the learning process over to the students. Active learner environments yield better results even though they are harder to manage than lecture-oriented approaches [3].

Cognitive research links student construction of new knowledge to students' pre-existing ideas about the world. In addition, there is a body of research that looks at how students' motivational beliefs and classroom contextual factors affect conceptual change [4].

It turns out that faculty background is not the only, and not even the most important, factor determining a learning outcome. A 2011 experiment at University of British Columbia [5] showed that

IOP Conf. Series: Journal of Physics: Conf. Series **1161** (2019) 012002 doi:10.1088/1742-6596/1161/1/012002

the structure of the instructional strategy has a significant impact on the learning outcomes. From the paper abstract: "We compared the amounts of learning achieved using two different instructional approaches under controlled conditions. We measured the learning of a specific set of topics and objectives when taught by 3 hours of traditional lecture given by an experienced highly rated instructor and 3 hours of instruction given by a trained but inexperienced instructor using instruction based on research in cognitive psychology and physics education. The comparison was made between two large sections (N=267 and N=271) of an undergraduate introductory physics course. We found increased student attendance, higher engagement, and more than twice the learning in the section taught using research-based instruction." The experimental section lessons were highly structured with pre-instruction assignments and interactive lesson activities that included clicker-questions-based discussion, group work and mini-lectures. While this approach may be non-intuitive to traditionally trained instructors, it is hard to argue with the data that show that the experimental section outperformed the traditionally taught section by 2.5 standard deviations.

In a 2014 article Scott Freeman and colleagues report on a meta-study involving 225 studies and 29000 students in various STEM discipline [6]. Some students attended lecture only classes, others were in various interactive settings. The authors report that interactive classes improved exam performance by half a standard deviation and lecture-only classes increased failure rate by 55%. Apart from obvious benefits to students, the authors point to some interesting broader issues. Based on conservative assumptions, the results of the studies translate into over US$3500000 in saved tuition dollars for the study population, had all students been exposed to active learning. If active learning were implemented widely, the total tuition dollars saved would be orders of magnitude larger. In another interesting side remark the authors point out that if the experiments analyzed here had been conducted as randomized controlled trials of medical interventions, they may have been stopped for benefit—meaning that enrolling patients in the control condition might be discontinued because the treatment being tested was clearly more beneficial.

In a 2009 article cognitive scientist Michelene Chi develops a hierarchy of learner activities that, ranging from active, through constructive to interactive, lead to better and better learning outcomes [7]. Being active can be characterized as doing something while studying. Another set of overt activities that students can perform while studying can be characterized as more constructive because in undertaking them, learners produce some additional outputs; and such outputs often contain new content-relevant ideas that go beyond the information given. Being interactive can refer to several types of overt activities, such as a learner talking with another person (who can be a peer, a teacher, a tutor, a parent) or responding to a system (such as an intelligent tutoring system, an animated agent). In the article Chi argues that extensive research shows that interactive activities lead to the best and most lasting learning outcomes.

Research recommends that learning units start with an examination, by the student, of his/her current knowledge regarding the topic to be studied, and with an examination of motivational beliefs that determine the student's approach to the topic. This approach is currently favored in any setting, but it is particularly appropriate when the audience is non-traditional students who need to be given some control over what they do to avoid slipping into surface learning which creates a dichotomy between their other (to them meaningful) activities and going to school (not to learn but to get certified for something.) We believe that carefully structured pre-instruction learning modules prepare students for a more productive time with a live teacher. The pre-instruction assignment informs the student and the instructor and sets the stage for a live lesson, which is then on a much higher level and closer to meeting the needs of the audience in the classroom than a traditional generic lecture.

In this paper, I describe two approaches that guide student through activities, which aim to achieve the goals outlined above. The first one, Just-in-Time Teaching [8], has been around for over twenty years. We introduced it in 1996 to take advantage of the internet. In the intervening years it has been adopted in all STEM disciplines and in many humanities disciplines [9]. The second initiative described in this paper, worked-examples, is work in progress that also shows much promise.

You are preparing an introductory biology lesson on cloning. You would love to face a class that has done some preliminary reading, has thought about the subject, and has some questions for you. Don't we all wish for that? Well, it can happen.

2. Just-in-Time-Teaching (JiTT)

When Kathy Marrs at is preparing her lesson on genetics she sends her students to the course website and invites them to ponder some warmup questions [10]. After some thought, students post their attempts at answers on the course site from where Kathy retrieves them just before class. She carefully weaves the student thoughts into her lesson.

A sample question: Dolly the sheep is a genetic twin, or clone, of a sheep that was born 6 years earlier than Dolly. Read the material for today. Does Dolly have parents? If so, who were her genetic parents? Is Dolly a "virgin birth"?

Two student answers:

A. That's a hard question. She was cloned from the cell of a 6-year-old sheep. Dolly doesn't have what we would commonly think of as parents, but genetically her parents would be the same parents of the six year old sheep. So, her parents would be the parents of the sheep she was cloned after.
B. Dolly does have genetic parents. Those being the parents of the sheep who was born six years earlier. Dolly would have the same genetic makeup as the sheep before, who has gotten her genetic makeup from her parents.

The Just-in-Time Teaching strategy pursues three major goals:

- To inform instructors about students' thinking. What do they know when they come to class? What have they learned from recently-assigned reading? What connections have they made based on this reading? And, how do students feel about course content?
- To structure the out-of-class time for maximum learning benefit by students.
- To provide information and content so that instructors can maximize the efficacy of the live classroom session.

Although Just-in-Time Teaching makes heavy use of the web, it is not to be confused with either distance learning nor with computer-aided instruction even though it is used in both of those settings. All JiTT instruction occurs in a classroom with human instructors. The web materials, added as a pedagogical resource, act as a communication tool and as an organizer. Since much of the learning occurs outside the classroom, JiTT practitioners view their pedagogical strategy as feedback loops between teaching and learning and between in-class and out-of-class experiences.

The JiTT classroom differs from the traditional lecture in two significant ways: First, having completed the web assignment just before class time, the students enter the classroom ready to participate actively in the activities. Secondly, students have a feeling of ownership because the interactive lesson is based on their own wording and understanding of the relevant issues.

JiTT is used in many disciplines at many levels. JiTT assignments vary significantly from discipline to discipline and from instructor to instructor.

A good JiTT question:

- Yields a rich set of student responses for classroom discussion.
- Requires an answer that cannot just easily be looked up.
- Encourages the student to examine his/her prior knowledge and experience.
- Requires that the student formulate and justify the response, including the underlying concepts, in his/her own words.
- Contains enough ambiguity to require the student to supply some additional information not explicitly given in the question. (This feature enriches the subsequent classroom discussion).

Here are eleven "good" JiTT questions:

1. Introductory Biology: What is the difference between a theory and a belief? You may want to look these terms up before answering. Be as specific as you can and give an example of each (Marrs).
2. Logic: The people of finiteland only tell the truth on Sunday, Tuesday, and Thursday. Which day of the week is it if a person from finiteland says, "I told the truth yesterday"? (Watt).
3. Introductory Physics: Please explain in your own words what a focal length is. Try not to use any equations or refer to a specific type of mirror or lens (Gavrin).
4. Introductory Earth Science: When reading or hearing news reports about dinosaur discoveries, what questions should you think about and ask yourself to evaluate the accuracy of the reports? (Guertin).
5. Advanced Mechanics: Your everyday experience, common sense, and math classes all suggest to you that the shortest distance between two points in two-dimensional space is a straight line (unless, of course, you want to consider a warped space-time continuum!). Please describe how you would go about proving that this is true, using the methods of variational calculus. Don't actually do this but describe in English the procedure you would follow. Please be specific in your steps. (Patterson).
6. Chemistry: Explain in simple terms how you solved warm-up #1. In particular, what questions did you ask yourself and what conclusions did you draw from the answers? (Blake).
7. Statistics: Estimate the probability that a North American male is precisely six feet tall. (Watt).
8. Introductory Physics: Let's say you have a prescription for contact lenses of -2 diopters. If you accidentally get glasses (as opposed to contacts) made to the same strength, your prescription will be a bit off, and you won't be able to focus at infinity. Estimate how far you will be able to focus. (Gavrin).
9. Earth Science: In your opinion, do we need to save ancient Egyptian monuments? Why? What are the monuments worth - are they worth anything? (Guertin).
10. Economics: Imagine that you are a newspaper reporter assigned to write a story about the upcoming meeting of the Federal Open Market Committee. Make a list of three questions that you will want to ask. For each question, explain carefully why it is important and what answer you expect from the Committee (Maier).
11. Introductory Biology: Allison is driving with her parents, Kate and Bob, when they get in a serious car accident. At the emergency room, her doctor (you) tells Allison that her mother is fine, but her father has lost a lot of blood and will need a blood transfusion. Allison volunteers to donate blood, and you tell her that her blood type is AB. Bob is type O. (a) Can Allison donate blood to Bob? Why or why not? (b) Allison, who is a biology student, begins to wonder if she is adopted. What would you tell her and why? (Marrs).

The questions illustrate some of the broad categories targeted by the pre-instruction assignments:

1. Preparing for a discussion of a complex, possibly controversial topic. Student classroom participation will be enhanced if they come in with informed opinions that they are eager to defend. Examples above: (1)(9) (11).
2. Creating a need to know. Good questions are sufficiently captivating so that even weak students may be interested in the answer. It is up to the instructor to build on this emerging motivation and sustain a lasting interest in the subject. Examples above: (2) (11).
3. Going beyond the bare definition. In many subjects, dry definitions of technical terms and complex concepts leave the students disinterested at best and often perplexed and confused. They then resort to coping strategy that disconnects the subject taught from the real-world experience. The physics community is well aware that erroneous perceptions of how the physical world works persists after successful completion of a physics course. Students function in two worlds, the everyday real world, and the “physics” world. It is very helpful to the student if he/she is challenged to digest the new enough to be able to construct his/her own formulation of the new ideas, crude as such formulation may be, to fuse the two worlds into one. “In your own words…” kinds of questions fall into this category. Examples above: (3)(5)(8).

4. Metacogniton. Extending the understanding. Ideally students would move from applying new the knowledge at the present level of understanding to a deeper synthesis between their current knowledge and the richer emerging knowledge. To do this they have to be encouraged to critically monitor their own learning and take advantage of the classroom session to check their progress. Examples above: (4)(6) (10).

3. Worked-examples

Worked-Examples are learning tools that provide students opportunities to study experts' approaches to problem solving [11]. The experts' solutions are paired with matched practice problems encouraging students to actively explain the examples to themselves. The worked-examples idea was proposed by Sweller whose research showed that traditional practice-based problem solving was less effective for beginning learners than guided study of fully worked out examples [12]. Sweller reasoned from the cognitive load theory perspective that when presented with traditional practice exercises students employ a trial-and-error or pattern-matching approach while students who analyze worked-out examples before being asked to work on their own tend to focus more on structural and conceptual issues.

The worked-examples approach to learning is of particular importance in the initial stages of skills acquisition as defined for example in the Adaptive Control of Thought--Rational model of cognitive architecture which holds that skills acquisition evolves over four overlapping stages. In the first stage learners solve problems by analogy, i.e. they refer to known examples when presented with a problem to solve. In the second stage they develop abstract conceptual schemas that guide their problem solving. After long practice they move to the third stage where the problem-solving process becomes rapid and automatic so that familiar aspects of the process do not consume attention resources. In the fourth stage, after having seen many diverse examples, they can often retrieve solutions directly from long-term memory. The value of the worked-examples approach is very high when the learners are in the first stage, beginning to transition into the second stage, e.g. students in introductory courses.

Worked examples work is paired with self-explanation effects developed by Chi [13] Observing students analyzing worked examples Chi noticed that when reaching an unexplained step, some learners paused and generated their own explanation of the justification for the step. Chi labelled this the self-explanation effect. In our work, we encourage students to self-explain by asking prompting questions.

Over the past twenty years, the worked-examples approach has been studied extensively, mostly in experimental settings, as part of the cognitive load theory research. Factors that moderate the pedagogical effectiveness of worked-examples include a) intra-example features; the intrinsic design of the example; how the solution is presented, b) inter-example features; a careful design of the relationships between multiple examples and c) the design of the prompts that lead the learner through the self-explanation process.

Research has identified three areas of importance when integrating the components: integration of diagrams and text to avoid the split-attention effect, simultaneous use of multiple procedures, such as text and simulations and, when appropriate, calling attention to the solution subgoals [14].

Inter-example features identified by research are the number of examples used per lesson, how the examples are varied and sequenced, how the example themes are used and how examples are intermingled. In general research shows that several examples (at least two) of different complexities, paired with practice problems seems to be the optimal configuration [15].

Self-explanation can be encouraged by structuring the example in such a way that the learner is prompted to self-explain, directly trained in self-explanation (the approach we take in this project) or encouraged to do cooperative group work. Self-explanation is particularly effective when the worked-examples are conceptually oriented and illustrate domain principles.

4. Conclusion

Just-in-Time Teaching Pedagogy, discussed in this article, shows wide variety from discipline to discipline [9] but they share the common feedback loop features described. As such, the JiTT instructional design promotes the three factors identified as the most important for success in college: time on task, student-instructor interaction, and student-student interaction. It is important to appreciate that the choice and implementation of a particular teaching method will impact student and faculty attitudes and motivation as well as learning outcomes.

Teachers using JiTT report a large spectrum of results, ranging from significant affective and cognitive gains to very negative student reactions, disillusionment and even regression in learning gains. We have much anecdotal evidence that in the best of cases, after adopting student-centered teaching approaches such as JiTT, many faculties change their teaching philosophies, sometimes in significant ways. The editors of the JiTT Across the Disciplines book comment that "...the most profound effect comes in changing the way the instructors think about their own teaching, and more important, the learning of their students." [9, p. 135]. Faculty members who once viewed their role as being "conveyors of knowledge" shift to becoming "facilitators of student learning." This developmental shift increases the efficacy of their subsequent teaching, and ultimately increases the levels of "deep learning" experienced by their students.

Successful implementation of the JiTT strategy depends critically on acceptance the students. If the online assignments are seen as an add-on to the course, to be completed perfunctorily in the shortest time possible, and then discussed brie y at the beginning of class, before the "real" lecture, the results will be disappointing. Students will resent the extra work and, in comparison with passive teaching and learning, no gains will be reported.

Acknowledgments
The author gratefully acknowledges the on-going support of collaborators at various institutions: Dr Kimberly de La Harpe and Dr. Brian Patterson at the Air Force Academy, Dr. Andy Gavrin at Indiana University Purdue University Indianapolis, Dr. Jeff Loats at the Metropolitan State University at Denver and Dr. Steve Spicklemire at The University of Indianapolis. Particular thanks are due Professor David Miranda at Universidad Industrial Santander for his continued interest and collaboration on this work. We are currently working on worked examples for advanced courses in physics. For examples please contact the author.

References

[1] Bransford J D, Brown A L, Cocking R R 2000 *How people learn: Brain, mind, experience, and school* Ed. National Research Council (Washington: National Academy Press)
[2] Bonwell C and Eison J 1991 *Active learning: Creating excitement in the classroom* Ed. Ashe-Eric Higher Education Report No. 1 (Washington: George Washington University) pp 2
[3] Hestenes D 1998 *Am. J. Phys.* **66(6)** 465-467
[4] Beghetto R A 2004*Practical Assessment, Research & Evaluation* **9(15)** 1-5
[5] Deslauriers L, Schelew E and Wieman C 2011 *Science* **332(6031)** 862-864
[6] Freeman S, Eddy S, Mcdonough M, Smith M K, Okoroafor N, Jordt H and Wenderoth M P 2014 *PNAS* **111(23)** 8410-8415
[7] Chi M T H 2009 *Top. Cogn. Sci.* **1** 73-105
[8] Novak G M 2011 *Just-in-Time Teaching* Ed. Buskist W and Groccia J E Evidence-based teaching. New Directions in Teaching and Learning (San Francisco: Jossey-Bass)
[9] Simkins S, and Maier M 2009 *Just-in-Time teaching across the disciplines and across the academy* (Virginia: Stylus Publishing)
[10] Marrs, K.A. & Novak, G.M. 2004 *Cell Bio. Edu. 3* 49-61
[11] Atkinson R K, Derry S J, Renkl A and Worthan D 2000 *Rev. Educ. Res.* **70(2)** 181-214.
[12] Sweller J, and Cooper G A 1985 *Cognition Instruct. 2***(1)** 59-89
[13] Chi M T H, and Bassok M 1989 Learning from examples via self-explanations In L B Resnick Ed. *Knowing, learning, and instruction: Essays in honor of Robert Glaser* pp. 251-282 (Hillsdale NJ: Erlbaum)
[14] Sweller J 1994 *Learn. Instr.* **4(2)** 295-312
[15] Reed S K, Bolstad C A 1991 *J. Exp. Psychol. Learn.* **17** 753-766

IOP Conf. Series: Journal of Physics: Conf. Series **1161** (2019) 012003 doi:10.1088/1742-6596/1161/1/012003

Articulation of regional headquarters in the Industrial University of Santander for a unified basic cycle of engineering

R F Valdivieso Bohórquez[1], **E L Castellanos-Leal**[1], **D A Miranda Mercado**[1], **J H Martínez Téllez**[1], **M J Sánchez Soledad**[1], **A R Lizcano Dallos**[1] **and G A Patiño Benavides**[2]

[1] Escuela de Física, Universidad Industrial de Santander, Bucaramanga, Colombia

[2] Vicerrectoría Académica, Universidad Industrial de Santander, Bucaramanga, Colombia

E-mail: rafvalbo@uis.edu.co, dalemir@uis.edu.co

Abstract. The Industrial University of Santander has five headquarters in Santander and one of the main challenges is to achieve articulation between subjects of the basic engineering cycle and the processes they entail. ExperTIC project, developed within the framework of the Academic Excellence System by Academic Vice-rectory of the University, who is supporting this articulation, standardizing the Teaching-Learning processes, creating and administering didactic resources in a centralized manner, facilitating access from all students to resources, through content management platforms such as expertic.uis.edu.co and Moodle virtual classrooms. In addition, information and content assets have been consolidated for Department of Physics, through links to e-books and Wikiversity pages. At regional headquarters of the Industrial University of Santander, evaluation is supported through the design and implementation of the evaluation process coordinated with a defined model, clearly establishing the stakeholders, using management tools; all this in order to able to design evaluations that will be applied simultaneously in all headquarters.

1. Introduction

The modern teaching process should place more emphasis on research on the Teaching-Learning process and less emphasis on knowledge and skills transmission through conferences. NASA, in a project, carried out with 10th and 11th grade students, has put at your disposal information, games, animations and other fun tools about environmental phenomena on its website, arguing the need for a paradigm shift in physics teaching, with the greatest scientific tool, "curiosity" [1,2]. One of the problems in high education is that content offered in many subjects don't really correspond to those legally approved in the academic agreements of the institutions, generating great confusion and demotivation in teachers and students.

Virtual resources and electronic learning systems have become common tools in many educational establishments around the world. TECH8 is a modularly designed model in Slovenia [3] and is based on collecting metadata and vital variables system for teaching process. This proposed system supports individualization and adapts to the knowledge level and understanding of each individual on the subject, which has generated better results than traditional teaching. In Industrial University of Santander (UIS) since 2015, a methodology for the realization of Physics 1, Physics 2 and Physics 3 laboratories (classical mechanics, electromagnetism and, waves and particles, respectively) based on research projects for the strengthening of the research competences of the engineering students have been developed. This methodology is based on several Teaching-Learning strategies, such as Just in Time Teaching (JiTT), active learning and mediated learning that together with a virtual platform allows students and teachers to access teaching resources and improve their experience.

In UIS' regional headquarters (UIS-RHQs) exist some difficulties in communicating with teachers has favored some problems, such as the lack of clarity in professors and students about requirements for programmed contents, contents readjustments at the will of the teachers for not having clear guidelines from each department, students with unequal opportunities in terms of available resources, few options for training teachers, non-standardized procedures for handling and disseminating summative activities and definitive qualifications, among others. In this sense, UIS has been working since 2016 with ExperTIC project, which seeks to establish processes for these subjects' articulation in RQHs, in order to provide students with digital resources such as: shared workshops, links to virtual books, Session Preparation Questionnaires (SPQ) and some actions to incorporate active learning strategies [4, 5, 6, 7]. This article presents some strategies for basic science cycle subjects articulation in UIS-RHQs, describing the methodology composed of the following processes; (i) model definition phase for the implementation and determination of the main processes, and (ii) implementation phase, where objectives socialization is carried out, training in pedagogical strategies for participants and the processes implementation are established.

2. Articulation model and methodology definition

In Teaching-Learning processes associated with subjects of the cycle basic sciences engineering offered at RHQs some problems were identified and, subsequently, solutions in Table 1 were proposed.

Table 1. Problems and solutions detected in the Teaching-Learning processes

Problems	Proposed solutions
• Teachers and students with little clarity about the scope of the contents. • Readjustments of contents at the will of the teachers.	• Declaration of detailed agendas based on academic agreements. • Facilitation of agendas for teachers and students. • Provision of follow-up matrices for the topics to be dealt with in each module.
• Students with unequal opportunities in terms of available resources.	• Creation of mechanisms to share workshops of all the professors with the students of all the headquarters.
• Few options for training teachers. • Non-standardized procedures for managing and disseminating summative activities and final qualifications	• Establishment of unified qualification matrices, in Moodle platform. • Teachers training in didactic strategies and in Teaching-Learning processes through "School of trainers". • Establishment of communication strategies for the deployment of guidelines from the Department Management. • Generation of virtual and face-to-face meeting sites so that teachers can share their experiences and challenges. • Regular visits to RHQs to communicate developments and raise findings of the Teaching-Learning processes.

2.1. Standardization of processes

In order to standardize processes, the *ExperTIC* project implemented the following strategies: Modular Content Manager (MCM), some teaching strategies for active learning and the definition of a regular conduit for teachers, which will be described below.

2.1.1. Modular content manager (MCM). It is the process that is carried out for the creation of intervened subjects' agendas; it is based on a model called Modular Content Manager (MCM) [8]. In this model used by the ExperTIC project, initially, a collection of the contents proposed by Departments and a comparison between them is made, starting from contents declared in Agreement 105 (2004) and Agreement 034 (2005) of the Academic Council. MCM is composed of the following elements, which are represented in Figure 1.

- Minimum content unit (MCU) or Content theme, it is each specific topic that integrates a subject and that is defined by Departments in their educational curriculums. In some cases, contents are very well detailed; while in others, it is necessary to decompose the topics.
- Minimum unit of learning (MUL) or Lesson, is a short cycle, with a defined duration that seeks to deal with content issues and that allows defining clear goals, objectives, and goals.
- Modules are units that cover one or more lessons and that generally use a summative evaluation for their closure. In a traditional course, Module is also known as a chapter or cut for an evaluation.

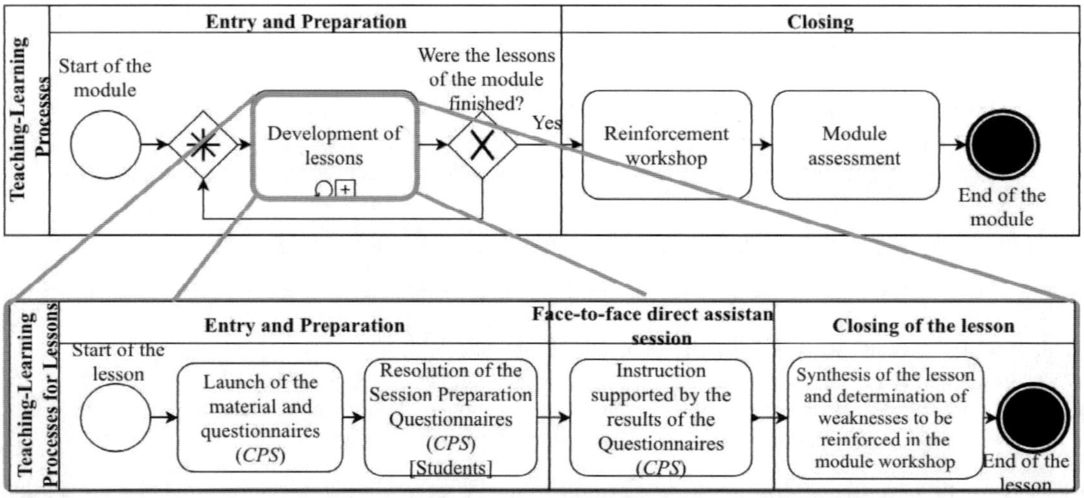

Figure 1. The life cycle of modules and life cycle of lessons that make up modules.

2.1.2. Adopted teaching strategies. The ExperTIC project uses two educational components (i) didactic strategies based on active learning [9], mediated learning [10] and Just in Time Teaching (JiTT) [11] and (ii) virtual learning management resources such as Moodle and the platform "expertic.uis.edu.co". These components were adapted to the basic cycle subjects (Calculus, Physics, and Chemistry, among others), impacting Learning habits of students in the previous preparation of their classes, generating a positive effect after it [12,13]. The previous study carried out by students in each subject, is supported with digital resources that are provided on the platform, among which are the Session Preparation Questionnaires (SPQ), links to digital books of the university's library and wikis to reinforce concepts, taking into account the modular organization that governs the course's content.

2.2. Provision of didactic tools for students

2.2.1. Creation of virtual digital tools. The virtual digital library Ebooks-7/24 provides all books, with links to each topic established in the MCM model of each subject. In addition, the SPQs associated with each lesson were created, focused on testing the basic concepts of each topic, which will be covered during the week, supported by the wiki of each of the lessons.

2.3. Coordinated evaluations and visits to the RHQs

For subjects' module evaluations in RHQs, a four levels maturity implementation model was established as shown in Table 2.

Table 2. Maturity model of evaluations by department.

Level	Characteristics
1	Independent evaluations. Teachers define dates of their evaluations before starting the course.
2	Coordinated assessments by department in each headquarter. Each headquarter independently coordinate a single evaluation per subject.
3	Coordinated assessments by department for different headquarters. Headquarters together coordinate a single evaluation per subject.
4	Coordinated assessments by department for all headquarters. All headquarters coordinate a single evaluation per subject (including main headquarter).

When levels two or more are implemented, coordination with the teachers is required to agree on the day and time of the evaluation, and content themes to be included in the evaluation, for which clear procedures are defined to help the logistics for these levels, as described in Figure 2.

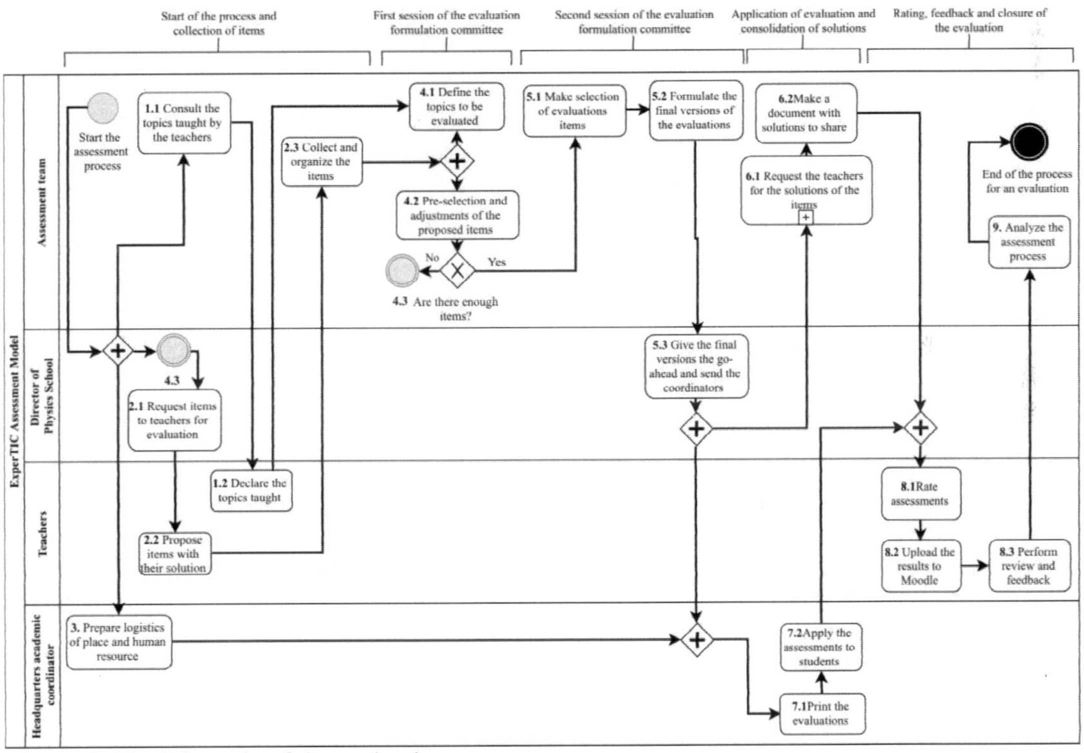

Figure 2. The life cycle of the evaluation process.

2.3.1. Implementation of the evaluation. Activities involved in the evaluation model have been readjusted since the implementation began. The process begins with the questions request to professors of each RHQs, taking into account the topics to be included in the evaluation that has been verified by teachers through a matrix for the follow-up of topics seen in class. Then, the evaluation team schedules two work sessions. The first session consists of a meeting by videoconference with each RHQs in which professors review and defend each question or problem previously proposed for the evaluation. The result of this meeting is the preselection of refined questions and with adjustments proposed by the professors. Later, in a face-to-face meeting (second session) the questions or definitive problems that will be included in the evaluations with their weighting are selected.

2.3.2. Visits to RHQs. Yearly, two visits to each RHQs were scheduled, with the purpose of documenting the experiences of the teachers and students in the use of the platform and methodology proposed in the project.

3. Results and discussion

3.1. Indicators of the ExperTIC project

Since 2017, the ExperTIC project has been implemented in the UIS-RHQs. The indicators in the last two years are summarized in Table 3.

Table 3. Goals achieved by the ExperTIC project.

Indicator	Goal achieved
Number of trained teachers in the training program on ICT strategies and methodologies	30
Number of student enrollments that use e-learning with the Moodle platform	4000
Number of subjects designed and implemented with virtual learning activities	12
Number of courses, with support in virtual classrooms of the Moodle platform	200
Number of accesses per year to library resources through links provided by ExperTIC	1000

In each academic semester, teachers of theoretical physics and laboratory have been incorporated in the training program organized by CEDEDUIS, through the training and continuous improvement of teaching, counseling pedagogical, methodological innovation and the use of information and communication technologies. The process started in 2015 with only Physics laboratories in the main headquarter. At present, there are 12 subjects (3 Physics, 3 Calculus, Differential Equations, 2 Linear Algebra, Analytical Mechanics and Chemistry).

3.2. Perception surveys of teachers and students

A 71.4% of the teachers belonging to the Bucaramanga, Barbosa and Socorro cities (14 in total) mentioned that the links to the books in each of the proposed topics in the MCM model are useful, a 21.5% consider them moderately useful and only 7.1% don't find much use. Regarding, the laboratory-theory relationship, the majority of teachers insist that topics covered in theory are not in accordance with the developed laboratories, because in many cases practices don't follow the order of the topics seen in class or there is a slight gap. On the other hand, they indicated that it is possible to plan a practice where the theoretical concepts of laboratory topics are not required and can be built with practice, and then correct and revise when it is treated in the theoretical classes. Finally, teachers agree that more communication channels should be established so that theory and laboratory teachers can interact a little more and not be seen as two independent courses.

On the other hand, some students from Physics 1 of Socorro city indicated that coordinated and organized by own teacher assessments have a similar degree of difficulty. However, it seems more appropriate that is done by the teacher, because sometimes some teachers don't reach to touch all topics proposed in the evaluation. Finally, Figure 3 shows the result of the general perception of students surveyed in Physics on a scale of 1 to 5 in the first semester of 2018.

Thirteen (13) students from Malaga and five (5) students from Socorro participated in the survey for Physics 2. Students consider that some improvements should be made to Moodle virtual classroom, among which are: a greater explanation and better quality of laboratories' videos and include presentations as a summary in each Lesson. These two previous aspects have been taken into account for the implementation of 2018 within the platform some old practices have been replaced by new ones with improved videos, and, review presentations have been added as wikis in a Wikiversity account, they summarize each of the topics by lesson, with the advantage that information can be modified, added and refined by Physics teachers. With respect to evaluations, as in Physics 1, students have different opinions, some affirm that it is appropriate to unify evaluations as proposed, others, however, consider that it is necessary to ensure that proposed topics in the exam, others insist that more time be given to respond and to improve the impressions' quality so that images of proposed problems are clearer.

Physics 3 students had a greater participation in the UIS-RHQs. Unlike the findings in Physics 1 and Physics 2, coordinated exams were considered more appropriate, because they perceive a similar level of complexity and some advantages such as equality when evaluating the knowledge acquired in the subject and the greatest responsibility in the course.

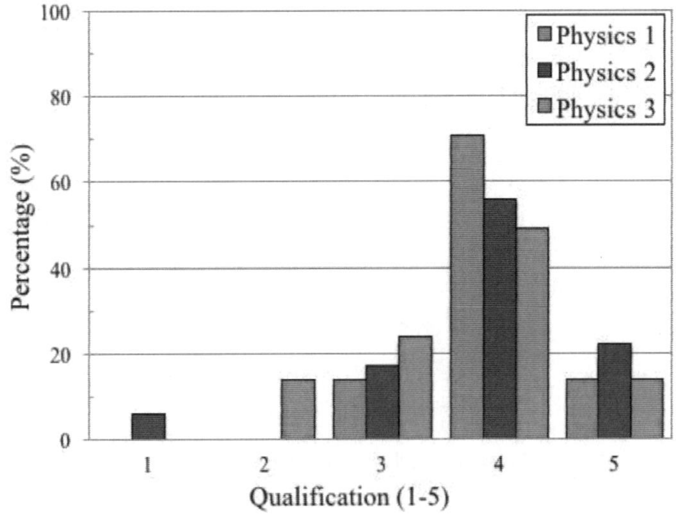

Figure 3. The answer to the question: Rate from 1 to 5 the general experience in Physics.

3.3. Findings found in the visits made to the Regional headquarters

During this first visit to Barbosa and Socorro cities, a couple of meetings were scheduled (one with students and later one with teachers). In this activity, Barbosa students contributed with suggestions and doubts regarding the implementation. Among the most important comments are: (i) evaluations have been mostly taken care by administrative staff and not by professors, which has generated problems when resolving doubts on specific topics, for which it is recommended look for other schedules in which teachers are present and (ii) extend the duration of the exam to 1.5 h to 2 h. At Socorro, teachers showed great interest in the project and proposed to make different points in the exam or mix the groups of different physics to avoid fraud and review contents of the accreditation program.

In Malaga, teachers and students expressed some concerns about the implementation, which were clarified. On the other hand, students affirmed that adaptation to evaluations coordinated has been complex, in a general way they have not had good results, they suggest that they should ask questions with a lower degree of difficulty. In the case of the platform and digital resources, as in the previous two cities, laboratories' videos improvement and the inclusion of review slides for Physics content are included.

Finally, teachers of basic engineering cycle are not very familiar with the proposed methodology in Barrancabermeja. However, some teachers showed some experiences where they have applied some didactic strategies in their courses, which can be articulated with Moodle platform tools and support the Teaching-Learning process. Furthermore, some teachers generally suggest forming "excellence committees" to consolidate communities, where strategies are proposed, continue to build content assets with experts support and link more tools for virtual content creation and educational resources open. Students welcomed the project explanation, didactic strategies that support it and the virtual tools created to support Teaching-Learning processes.

IOP Conf. Series: Journal of Physics: Conf. Series **1161** (2019) 012003 doi:10.1088/1742-6596/1161/1/012003

4. Conclusions

From the visits made to UIS-RHQs, it can be concluded that students and professors have very well received the implementation of the ExperTIC project, although some concerns persist associated with the fact of changing study habits and becoming familiar with the coordinated exams. On the other hand, new teachers of subjects such as Calculus and Chemistry are motivated by the experience lived in Physics, but they need more accompaniment to have clear instructions.

Among the aspects that should be strengthened in the project, the need to create a collaborative space in the virtual classrooms of Physics subjects was highlighted, where a summary of each Module is presented, with the possibility of that teachers can edit the information and improve continuously. To solve this request, the educational platform "Wikiversity" was used, which has been well received by teachers and students. On the other hand, one space was created within the Moodle platform to report errors and thus identify faults in the incorporated materials and quickly solve them.

Acknowledgments

The implementation of the ExperTIC project in the RHQs is due to the support provided by UIS through Academic Vice-rector and Department of Physics. Likewise, an administrative and logistical collaboration by Academic Subdirection of iPred Institute and the academic coordinators of each UIS-RHQs. Finally, a very special thanks to teachers and students have taken part in this project.

References

[1] Stoica I, Moraru S and Miron C 2010 *Procedia Social and Behavioral Sciences* **2** 3716

[2] Castro A and Santos L 2010 *IFIP Conference on Information Technology in educational Management* **348** (Berlin, Heidelberg: Springer) p 75

[3] Dolenc K and Abersek B 2015 *Computers & Education* **82** 354

[4] Miranda D A, Sanchez M J and Forero O M 2017 *IOP Conf. Series: Journal of Physics: Conf. Series* **850** 012015

[5] Zabala G 2007 *Ciencia Conocimiento Tecnologia* **52** 29

[6] Tristancho J A, Contreras L E and Vargas L F 2014 *Educación y desarrollo social* **8** 28

[7] Oliver-Hoyo M T, Alconchel F and Pinto G 2012 *Revista Española de Física* **26** 45

[8] Valdivieso R F 2017 *Modelado del Proceso de Docencia y Discencia de Enseñanza-Aprendizaje en Educación Superior* (Bucaramanga: Universidad Industrial de Santander)

[9] Freeman S, Eddy S L, McDonough M, Smith M K, Okoroafor N, Jordt H and Wenderoth M P 2014 *PNAS* **111** 8410

[10] Hakkarainen K 2009 *Computer-Supported Collaborative Learning* **4** 213

[11] Novak G 1999 *Just-in-Time Teaching: Blending Active Learning with Web Technology* (USA: Addison Wesley)

[12] Guertin A S, Zappe E and Kim H 2007 *J. Sci. Edu. and Tech.* **16(6)** 507

[13] Formica S P, Easley J L and Spraker M C 2010 *Phys. Rev. Spec. Top. – Phys. Edu. Res.* **6** 020106

IOP Conf. Series: Journal of Physics: Conf. Series **1161** (2019) 012004 doi:10.1088/1742-6596/1161/1/012004

Virtual tool for training and evaluation of problem-based engineering subjects

O A González-Estrada[1], J F Cáceres-Ramírez[2] and W Pinto-Hernández[2]
[1] GIC, Universidad Industrial de Santander, Bucaramanga, Colombia
[2] GIEMA, Universidad Industrial de Santander, Bucaramanga, Colombia

E-mail: agonzale@uis.edu.co

Abstract. In Educational Technology, one of the challenges in the use of virtual tools for engineering education is the assessment of problem-based subjects. In this work, a software generator of exercises is developed, to be contained in a virtual platform, which seeks to improve the learning process and the evaluation of the subjects based on problems such as, e.g., Mechanics of Materials. The student can use it for training, with content corresponding to the subject of the course, monitoring their performance and obtaining effective feedback. Additionally, the teacher has the possibility to periodically conduct formative assessments that allow him to recognize the concepts in which the students are failing. This tool can provide an objective instrument to normalize grading across groups directed by different lecturers. The software was written in the programming languages Java, JavaScript and HTML for the presentation of the contents, and can be used from any electronic device. This tool can be used to enhance classroom learning in face to face, blended or online learning.

1. Introduction

Since its inception, the World Wide Web has undergone several transformations thanks to technological advances and the needs that different types of users demand. Web 2.0 is the conception of the network as a platform, it gives people an active role, and encourage participation and usability [1]. These features have promoted the boom of Educational Technology aimed to enhance the teacher-student experience, improve knowledge management [2], and facilitate the learning process [3]. Within Educational Technology, electronic learning (e-learning) consists in the use of new multimedia and Internet technologies to improve the quality of learning by facilitating access to resources and services, as well as to the collaboration and remote exchange. In e-learning, the formative process takes place through a kind of virtual environment in which the teacher-student interaction takes place as well as the activities of the students with the learning materials. The e-learning serves as a means for face-to-face, online or blended teaching.

Educational Technology has many advantages, it can be used to simplify and reduce the cost for labs used to demonstrate concepts [4], massify access to knowledge with initiatives as massive online open courses (MOOC), improve feedback for effective formative assessment [5], or even help vulnerable populations by facilitating access to remote places affected by security issues [6]. Challenges for developing countries include improving Information and Communication Technologies (ICT) infrastructure, facilitating access and use of technological resources, and encouraging the adoption of new training methods [7].

We can find numerous efforts that have been reported in the literature to support the teaching-learning process of various subjects [8,9]. In [10] a program created in Macromedia Flash v8.0 focuses on the understanding of the laboratory tests that accompany the theory of the course Mechanics of Materials. It includes tests of traction, compression, shear, Poisson's effect, torsion, extensometry, bending, columns and photo-elasticity. The student has different types of evaluations that not only allow him to evaluate his learning but also prepare him for the tests that are normally carried out in theory courses.

In [11], a didactic tool is developed with the objective of supporting the teaching-learning process of Linear Algebra in engineering degrees. When entering the problem data in the software, the operation is processed and the procedure is visualized, showing step by step how the problem is solved. Moreover, it contains theoretical material to review the concepts that the student needs to develop the exercises. The proposed system was implemented in the C ++ Builder.

SimulPhysics [12] is a multiplatform software, written in Java, that contains a set of interactive simulations on different topics of physics that supports the assimilation of the concepts developed in the theoretical part of the course. On the platform, we can find simulations on different topics that are seen in the subject of Basic Physics, e.g., mechanics, optics, waves, etc.

Celosías paramétricas [13] is an Android application, targeted to Education and Learning, that automatically generates trusses by combining parameters related to the geometry, material, supports and loads, and shows the influence of each of them on its structural behaviour: reaction forces, stresses, and deformations. It allows the optimization of the structural design by determining the values that produce the lowest stresses and deformations for the highest loads, while minimizing the weight.

MDSolids [14] is an educational software package dedicated to the introduction of the Mechanics of Materials course. The software has an intuitive interface that guides the user to understand important facts that are found in different types of problems. It helps to visualize the nature of internal stresses and strains.

Regarding assessment, Hernández Pina [15] investigates the learning process in the university context, and concludes by stating that the way in which the teacher raises the evaluation affects learning approaches (superficial or deep) and the quality of such learning. Quantitative evaluation strategies lead to superficial learning approaches, while formative and qualitative strategies can produce deep learning and high-performance approaches [16]. In [17], a project carried out by the School of Engineering of the EAFIT University implements an interactive software in a virtual platform. With this tool the students can carry out tests, clarifying doubts and preparing them for the final test. It incorporates a rating system and a problem generator algorithm for the course of Statics.

Following the development of the new Information and Communication Technologies (ICTs), we propose a software tool for a virtual educational environment which could be used to improve the assessment of problem-based subjects and provide a standardized marking system that could be used across classes and campuses. The software generates problems on the different topics of the subject Mechanics of Materials, giving the student an interactive tool that allows him to learn by solving problems that help him to strengthen his analytical skills and increase his cognitive abilities. Moreover, the teacher could perform controls and follow performance indicators, which can be used as a reference of the level of study and understanding of their students on the content seen in class, and the competencies defined in the curriculum.

The paper is organized as follows: in the next section, we exposed the methodology, including the knowledge base and performance indicators, which are used to define the constraints for the design of the tool. In section 3, we present details related to the implementation of the tool in the Learning Management System. Conclusions are given in the final section.

2. Methods

The methodology is focused on the subject Mechanics of Materials, although it could be easily extended to other problem-based subjects in engineering curricula. The project is based on the framework of Educational Technology, which refers to the systematic way of conceiving, applying and evaluating the set of teaching and learning processes, considering both technical and human resources and the interactions between them, as a way to obtain a more effective education.

In this regard, we define a strategy to tackle different issues detected during the learning and assessment process in the schools of engineering at Universidad Industrial de Santander. First, when evaluating problem-based subjects, the teacher should pay attention to the process followed by the student to reach the answer. Normally, the problem solution is divided into steps, and a weighting is given to each of the partial steps. However, if the student fails early in the process, the teacher should evaluate the remaining steps considering the wrong values obtained at the beginning, and their impact in the final answer. This is particularly difficult for large groups, which finally yields non-homogeneous, subjective marks. This difference is accentuated if we consider groups led by different teachers. Moreover, regarding teachers, there is a notable difference in the assessment performed by tenured and non-tenured professors, mostly due to the correlation between the grades and the teacher evaluation, which affects contract extensions for non-tenured professors. There are even instances where the content given by the teacher is incomplete and we end up with groups of students with different level of competences reached. This non-standardized marking system gives rise to problems for the mobility of students across campuses and helps to increase retention rates.

We start by defining the knowledge base that encompasses the area of Mechanics of Materials, putting special attention on the competencies required at the end of the course. Next, we work on the performance indicators for the established competences, which help to define the constraints for the design of the tool.

2.1. Subject description and knowledge by competences

The subject Mechanics of Materials [18] studies the mechanics of deformable solids by simplified linear elastic models, relating in the design process the parameters of geometry, forces, and material. The mechanical engineer works with static and moving mechanical structures and systems. The design and calculation of these systems involve considering all the loads that act on each of its elements and the intensity of internal force (stresses) generated by these loads. In mechanical design, the stresses can be related to the mechanical strength of the material to assess the possibility of failure for a given element. In general, a machine element can fail because the stresses exceed the strength of the material, or because excessive deformation is occurring that prevents its normal operation [19]. The calculation of these stresses and deformations involves considering the internal forces and the geometric features of the element.

At the end of the course the student is expected to:

- Recognize the type of stress produced by a force or moment applied to a machine or structural element, according to the way it acts on the element.
- Calculate stresses and deformations that occur at any point of an element subjected to a specific type of load.
- Select the critical point where the highest stress occurs in any element subjected to several types of load simultaneously.
- Define the plane stress state at the critical point of any element subjected to different types of external load.
- Analyze the possible failure of an element subjected to different types of loads using different criteria according to the type of material and propose changes to the geometry accordingly.

The curriculum is divided into four fundamental thematic blocks. The first block addresses the introduction to the concepts of stress and strain, and the state of axial loading. The second block deals with elements subjected to pure torsion. The third block describes pure bending, typical of beams. The fourth block refers to multiaxial stress, combined load, pressure vessels and design failure criteria.

2.2. Performance indicators

The performance of the student is evaluated with quantifiable indicators, related to the competencies, which provide a global assessment of his understanding of the subject. These are:

- Identifies the type of load produced by a force or a moment applied on a machine or structural element and evaluates the corresponding stress and deformation.
- Recognizes the differences between the different types of stresses and deformations that appear in elastic problems.
- Identifies the critical point where the highest stresses occur in any mechanical element subjected to several types of load simultaneously.
- Calculates the principal stresses, maximum shear stress and equivalent normal stress at the critical point of any element subjected to different types of external load, according to a specified failure theory.

2.3. Design of the tool

In this section, the thematic and evaluative content of the course is identified, as well as the parameters that can be varied to generate multiple versions of the same exercises without modifying their knowledge structure. To carry out this process the following steps were followed:

2.3.1. Definition of the learning objective. To be able to define the exercise that is going to be proposed, we must know what we want to achieve with it, i.e., identify the cognitive competences that the student must acquire during the course. Here, we refer back to the competences listed in section 2.1

2.3.2. Definition of the exercises. A review of the curriculum of the subject Mechanics of Materials must be done, and the main topics of the subject are identified: axial loading, bending, torsion and combined load. After considering the thematic content, the bibliography presented by the teachers is reviewed in order to look for the most representative exercises. For each topic, a set of exercises is proposed in order to cover the main knowledge components of the cognitive competences.

2.3.3. Solution and generalization of the exercise. The selected exercises must be solved to create a base algorithm that follows the steps taken to obtain the answer, in order to generate several versions of the same problem by modifying certain parameters but preserving the knowledge component. We define and bound the parameters that can vary to enforce randomness in the data of the problem. In Table 1 we can see an example of the design of an exercise that was included in the software database.

The problem template in Table 1 considers the following points: (*i*) the topic and objective of the exercise describing the knowledge component, (*ii*) the problem statement where the questions to be answered are proposed, (*iii*) the list of parameters acting as variables of the problem, i.e., the problem data randomly generated by the software that must have a previously defined range to avoid errors in the answers, (*iv*) a general solution for the exercise is developed, which allows to code the exercise easily.

3. Software implementation

There are many advantages of developing the program as web software. It is multiplatform, and the student can use it from any electronic device. This approach is convenient as it makes the tool accessible from virtually anywhere, in agreement with mobile learning (*m-learning*) statement of students using mobile devices at their time convenience. The tool allows the teacher to use it to do quizzes and workshops. Any student will be able to propose new exercises, facilitating communication with the teacher and feeding the database to increase the knowledge base.

Due to the requirements of the software, especially the one focusing on facilitating the process for users with poor programming experience, there is a need to use an expression analyzer. An expression

IOP Conf. Series: Journal of Physics: Conf. Series **1161** (2019) 012004 doi:10.1088/1742-6596/1161/1/012004

analyzer is an algorithm that interprets a mathematical expression with pre-established functions, assigns values to each of the variables that intervene in the expression, and gives a numerical answer, see Figure 1. This condition speeds up the task of introducing new exercises to the database since instead of programming the algorithm, the existing one can be quickly modified to the required needs. Thus, the expression analyzer is used to separate the coding of the exercises from the main routines, without it, the software was conditioned to a person making a particular code for each of the exercises.

Table 1. Problem template of an exercise for axial loading.

Topic: Axial loading.

Objective: Determination of normal stress in the two-forces structural elements.

Problem statement:
There is a machine consisting of two tongs that support a cylindrical element, as shown in the figure.

- Determine the normal stress developed in the link AB.
- Determine the shear stress developed on pin A, consider that it is subjected to double shear.

Note: Units are given in international system, SI.

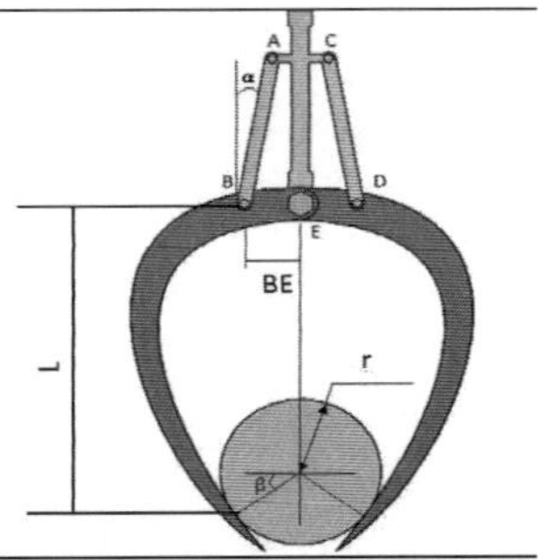

Parameters	Bounds	
	Max.	Min.
Cylinder weight [N]	1000	5000
Distance BE [m]	0.3	0.9
Distance L [m]	1	2
Cylinder radius [m]	0.1	0.5
Alpha	10°	80°
Betha	10°	80°
Diameter of the bar AB [m]	0.1	0.35
Diameter of pins [m]	0.01	0.08

Solution

FBD Cylinder	FBD Tong	Stress

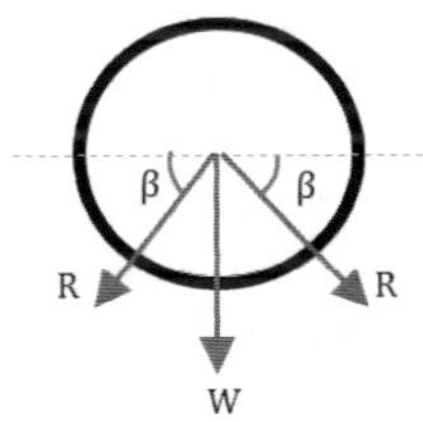

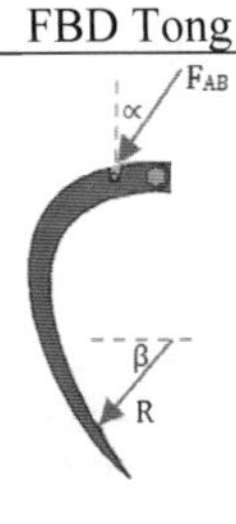

$$\sigma = \frac{F_{AB}}{A_{AB}}$$

$$\tau = \frac{V}{A_{pin}} = \frac{F_{AB}}{2A_{pin}}$$

$$\Sigma F_y = 0$$

$$R = \frac{W}{2\sin\beta}$$

$$\Sigma M_E = 0$$

$$F_{AB} = \frac{RL\cos\beta - (R\sin\beta)}{BE\cos\alpha}$$

After performing a search on the web, we found a previously coded expression analyzer with Lesser General Public License (LGPL) [20]. This analyzer allows simplifying the template definition for the proposed problems. Equations obtained during the solution steps are introduced in the template to parametrize the solution of the problem with random input.

The software was implemented in the proprietary platform COMA (http://calumet.github.io/) [21] used as the Learning Management System (LMS) at the School of Engineering. The software was codified in the programming languages Java and JavaScript, to follow a multiplatform (Linux, Windows, Android, Mac) approach. In addition, the HTML language was used for the presentation of

IOP Conf. Series: Journal of Physics: Conf. Series **1161** (2019) 012004 doi:10.1088/1742-6596/1161/1/012004

the contents. The database of exercises can be constantly updated since anyone with access to the platform can propose problems. The database contains different types of questions, i.e., multiple choice, calculated, true/false, matching, and essay. As indicated before, the focus is on the development of calculated questions for problems centred in the process. One key feature that is proposed in this implementation is that values are recalculated with the input values given in previous steps during the process of solving the problem. This segmentation of the process allows assessing the competences of the student independently. For the specific case of problems related to Mechanics of Materials, it evaluates the statics, the geometry, the stress and the strains calculations, all in different steps, regardless of whether the student failed a previous step.

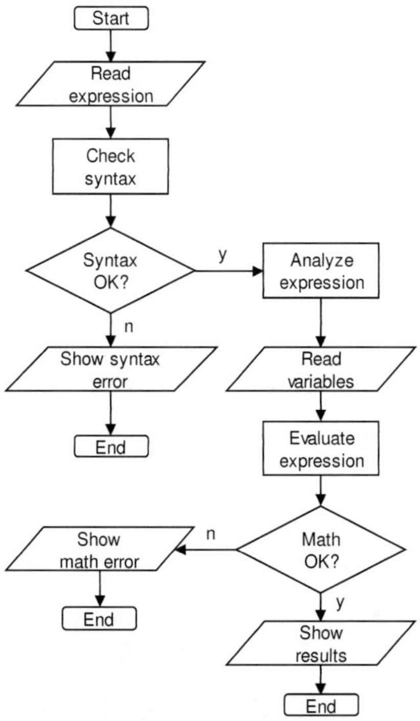

Figure 1. Basic algorithm of the expression analyzer.

The software offers two services: (*i*) *Training*: The teacher creates a workshop by setting the number of problems and the stipulated time to develop them. As the software has the option to review answers, the student can solve the exercise as many times as he wants, monitoring their results, until he is sure the procedure is correct. This improved feedback is required for effective formative assessment [16], and it is extremely difficult with mainstream education, (*ii*) *Evaluation*: The teacher previously creates an exam, setting a number of problems and an adequate time to solve them, and takes the students to a computer room in order for the students to carry out the activity.

Figure 2 shows a flow diagram with the internal process that is carried out when the teacher introduces a query of the type "create questions with expressions" in the system. For a teacher to include a question in the software, he must add the statement and the problem image, followed by the data and the mathematical equations to develop the exercise, then he chooses which expressions will be used as answer. Figure 3 shows snippets for the definitions of variables and expressions in the code.

When the student works on a workshop and presses the review option, the program follows a series of internal steps to show him if the answers are correct. This sequence consists in taking the randomly generated data and calculating the solution to the defined mathematical equations. After that, the software compares the values of the expressions indicated as answers with the values entered by the user, showing in red the boxes with wrong answers and in green the boxes with correct answers.

IOP Conf. Series: Journal of Physics: Conf. Series **1161** (2019) 012004 doi:10.1088/1742-6596/1161/1/012004

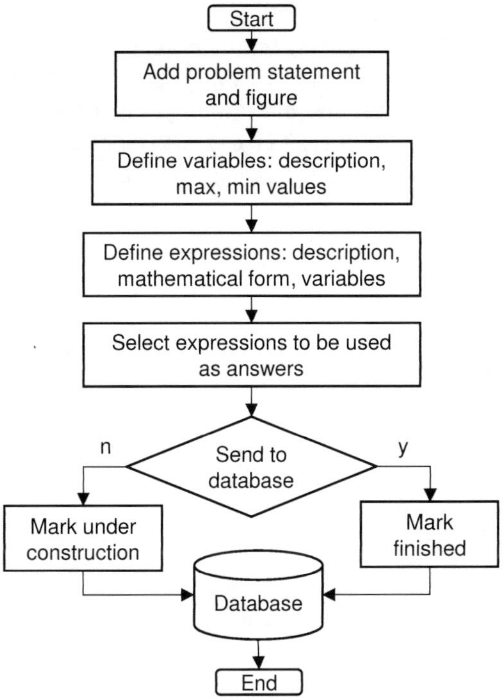

Figure 2. Basic algorithm for "create questions with expressions".

```
[{
  "type": "variable",
  "variable": "b",
  "description": "L   [m]",
  "min": "1.5",
  "max": "2.8",
  "value": ""
 },
...
 {
  "type": "variable",
  "variable": "z",
  "description": "Area bar AB   [m]",
  "min": "0.001",
  "max": "0.002",
  "value": ""
 },
...
```

```
 {
 "type": "expresion",
 "variable": "g",
 "expression": "a/(2*sen(p*e/180))",
 "description": "R",
 "answer": false,
 "value": ""
 },
...
{
 "type": "expression",
 "variable": "l",
 "expression": "k/z",
 "description": "Admissible shear stress AB",
 "answer": true,
 "value": ""
}]
```

Figure 3. Snippets of the array definition in JSON for the problem in Table 1, for variables and expressions.

4. Conclusions

We presented the development of a multiplatform software tool to be used for the training and assessment of students of the subject Mechanics of Materials. The tool, implemented in Java, JavaScript and HTML, randomly generates exercises based on a parametric definition of the problems. The solution is calculated and compared with the answer given by the students and can be used to train or mark their performance. The problems are segmented into steps based on the knowledge base and the competencies to be evaluated. The software is initially focused on the subject Mechanics of Materials, but it is easily extensible to other problem-based engineering subjects. By reviewing the answers, the student can check their errors and repeat the process to obtain effective feedback in the formative assessment process. This tool is based on innovative teaching strategies that seek to favor a teaching based on the active participation of the student. Moreover, it provides a standardized marking system that could be used across campuses and groups, reducing heterogeneities for large numbers of students due to subjective marking and non-uniform content.

Acknowledgements
We acknowledge the support given by the group Calumet, Alejandro Riaño, and Luis Ignacio González Ramírez.

References
[1] DiNucci D 1999 Fragmented Future *Print* **54** 32
[2] Lache L, León A P, Bravo E, Becerra L E and Forero D 2016 Las tecnologías de información y comunicación como prácticas de referencia en la gestión de conocimiento : una revisión sistemática de la literatura *Rev. UIS Ing.* **15** 27–40
[3] Moreno-Chaustre J J, Andadre-Sosa H H, García-González J R, Hernández-Pino U, Maestre-Góngora G P and López-Mólina G 2014 Modelo de evaluación para valorar el cambio en las prácticas docentes con TIC *Rev. UIS Ing.* **13** 7–22
[4] Machado-Mercado D, Herrera-Murgas G, Roldán-Mckinley J and Díaz-González J 2015 Una herramienta computacional didáctica para el análisis cinemático de mecanismos planos de cuatro barras *Rev. UIS Ing.* **14** 59–69
[5] Gikandi J W, Morrow D and Davis N E 2011 Online formative assessment in higher education: A review of the literature *Comput. Educ.* **57** 2333–51
[6] González-Zabala M P and Galvis-Lista E A 2014 Propuesta de un sistema de información para apoyar la operación de modelos de educación en situación de emergencia : caso círculos de aprendizaje *Rev. UIS Ing.* **13** 39–47
[7] Douglas J, Imran A and Turner T 2015 Designing an ELearning Portal for Developing Countries: An Action Design Approach *Australasian Conference on Information Systems* (Adelaide)
[8] Gallardo-Pérez H de J 2014 El modelo pedagógico dialógico crítico en la educación *Respuestas* **19** 81–92
[9] Rua Ramirez E, Barrera Siabato A and Moreno López N M 2014 Aprendizaje interactivo de termodinámica de fluidos apoyado en las tecnologías de la información y comunicación *Respuestas* **19** 41–50
[10] Torres J 2007 Software para la Enseñanza Didáctica de Mecánica de Materiales *5th International Latin American and Caribbean Conference for Engineering Technology* (Tampico: LACCEI) pp 1–7
[11] Almaguel Guerra A, Alvarez Mora D, Pernía Nieves L A, Mota Pimentel G J and Coello León C 2016 Software educativo para el trabajo con matrices *Rev. Digit. Matemática, Educ. e Internet* **16** 1–12
[12] Aristizábal D, Restrepo R, Ramírez C, Montoya N, González E and Muñoz T 2013 Uso de las NTIC para apoyar la enseñanza de la física básica para ingenieros: experiencia en la Universidad Nacional de Colombia sede Medellín *XIV Encuentro Internacional Virtual Educa* (Medellín: Virtual Educa) pp 1–19
[13] Mansilla F H 2014 *Celosías Paramétricas* (Madrid: Google Play)
[14] Philpot T A 2013 *MDSolids* (Rolla, MO: Missouri S&T)
[15] Hernández Pina F 1996 La evaluación de los alumnos en el contexto de la evaluación de la calidad de las universidades *Rev. Investig. Educ.* **14** 25–50
[16] Sadler D R 1989 Formative assessment and the design of instructional systems *Instr. Sci.* **18** 119–44
[17] Restrepo Ochoa J L 2012 Generador automático de tareas como apoyo a los procesos de evaluación *Reunión Nacional ACOFI* (Medellín, Colombia: ACOFI)
[18] Beer F P, Johnston E R J, DeWolf J T and Mazurek D F 2014 *Mechanics of Materials* (New York: McGraw-Hill Education)
[19] Ayestarán A, Graciano C and González-Estrada O A 2017 Resistencia de vigas esbeltas de acero inoxidable bajo cargas concentradas mediante análisis por elementos finitos *Rev. UIS Ing.* **16** 61–70
[20] Moreno Parra R A 2013 *Evaluador de Expresiones Algebraicas II* (Cali: Moreno Parra, Rafael Alberto)
[21] Calumet 2018 *COMA* (Bucaramanga: Calumet, UIS)

IOP Conf. Series: Journal of Physics: Conf. Series **1161** (2019) 012005 doi:10.1088/1742-6596/1161/1/012005

Review of the possibility of knowledge in educational sciences

J L Vera-Rivera[1]
[1] Instituto de Educación y Pedagogía, Universidad del Valle, Cali, Colombia.

E-mail: vera.jose@correounivalle.edu.com.co

Abstract: There are many trends that determine all the possibilities that exist in the education sciences. These are oriented to the formation of the human being. The current development is demarcated by the use of active methodologies through science, research, technology and innovation. where the Universities advance in the professional formation of the students through the virtualization of the education. In the present work we intend to review the current state of the research work at national and international level in relation to the different ways of covering the teaching - learning processes. For that, an exhaustive review of the bibliographical sources was made through the use of several specialized databases such as Ebsco, ISI Web of science, Latindex Open Access, Scielo and Scopus. The articles classified as most relevant by these databases were selected taking into account their citation index. As a result you can see a variety of possibilities that still use the traditional forms of education and citizenship training based on traditional master classes where the teacher is the protagonist of academic activity and you can also see new methodologies that apply the new possibilities of education based on virtualization and independent training based on a more globalized and participatory education where the student is the protagonist of their learning.

1. Introduction

This paper is the product of an analysis of the literature review that summarizes the subjects taught in the doctoral doctorate in education at the Libertador Pedagogical University in the Rubio municipality of the Bolivarian Republic of Venezuela. They also contribute the opinions and concepts of the topics debated and defended in the studies carried out by the author in the physical culture sciences at the Manuel Fajardo High School in Havana Cuba.

Education is a topic of debate and in which there is a great variety of researchers who contribute the contemporary tendencies in which the different teaching - learning processes are supported as an example we can mention the work of Jarpa [1]. In his essay he presents an example of equity with the concept of differentiated education to nurture the analysis of the phenomenon of First Generation students in higher education. To substantiate his thesis presents a good variety of background on access to higher education, also reflects on the concept of equity in this educational level and conceptually requires differentiated education as a strategy to include in the university classrooms as a mechanism of student retention. Also, the relationship between educational segmentation, equity and the possible implementation of differentiated education in university classrooms is discussed from a critical perspective. From this same study can be taken up Some Latin American studies: De los Ríos y Canales [2], Castaño et al. [3], Fernández [4], Acuña [5], Bolívar [6], Rojas, et al. [7], link the partial or total abandonment of a university career to academic reasons such as the deficit in reading and writing or socioeconomic factors. Specifically, it is recognized that a combination of academic, family and economic responsibilities that would prevent them from integrating studies and work, precipitating their early abandonment, operates in the most vulnerable juvenile subjects.

Therefore, it can be partially concluded that: There is a great variety of knowledge and writings that speak to us about education as a science and we can appreciate many epistemological forms and positions that help us to reflect on this. Thus, knowledge is an uninterrupted process, framed always in historical spaces and times, which allows discovering the facts and natural and social phenomena, and leads to the creation of a certain picture of the universe, unique to each individual individuality. In fact, people make reality their own in a very special and personalized way, according to their own needs and potential, and according to the circumstances of their social, cultural and educational environment, since they are always and, in situation, which they know from a context and a time, from history and from their own history.

2. Method

The methodological design used in this research work was an analytical study where the most important methods of pedagogical research are used through an essay that reflects the opinion of outstanding researchers in the area of education and pedagogy. The Figure 1 represents the research model.

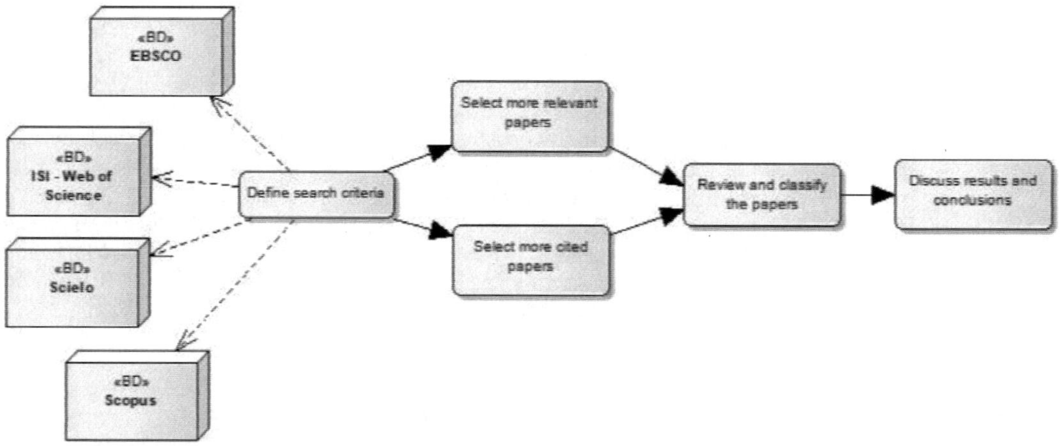

Figure 1. Research model.

A bibliographic search was carried out in the databases of the most important scientific publications such as Ebsco, ISI web of science, Scielo and Scopus, different search criteria related to the interests of this research were defined using criteria and words keys related to the object of study such as history of pedagogy, didactics of education, teaching - learning processes and virtual education all this to analyze the different possibilities of knowledge that there is in the sciences of education. To analyze and collect the information, the search criteria were defined in each of the databases, then the most relevant and the most cited articles were selected for each criterion, later they were ordered, and the most significant contributions are reported below.

3. Results. Possibilities of Knowledge in the education sciences

Becerra in his article describes a line of research on the history of Colombian pedagogical thought and publishes a wide variety of lectures, essays, on local and regional education. The pedagogical thought, in the present proposal of analysis, can be defined as the set of ideas by means of which it is thought about the education in which a set of representations is at stake by a group of intellectuals of the education, which they write about the school and the teaching-learning processes [8].

For Baudrillard, he proposes that the educational is "to do, a practice, an activity, which, intentionally or not, shapes man, but from there to argue that pedagogy is an activity, that is, an art, there is a great distance, because Pedagogy does not it is education, but the discipline that has it as an object "[9].

IOP Conf. Series: Journal of Physics: Conf. Series **1161** (2019) 012005 doi:10.1088/1742-6596/1161/1/012005

We can say that, for this author, Pedagogy is the only science of education. In the opinion of Herrera [10] defends that education cannot only, but should be subject to scientific knowledge, stating that many problems in the educational field, it cannot be treated empirically despite being conditioning educational action is not conceived an educational process that is not based on knowledge generated from the purely speculative. It would be a systematic and scientific process is not possible to understand in a task as complex as it is the educational, which act factors of biological, psychological, sociological, economic, etc, which need to be subjected to analysis of scientific order to determine as much as possible, not the totality, because we have already said that the behavioral sciences always maintain great difficulties. How they interact or influence and ultimately intervene in the educational process.

Pedagogy has to assume this problem of ends because from it, essential problems of the process are addressed in that search for perfection of man, which is the object of education and as Bolívar affirms beyond Applied pedagogy is the act of education, which is the putting into action by each educator of the knowledge of the doctrine and formulas taught by pedagogy [11].

Pedagogy as a set of sequences in which they are represented by this order: science, moral or practical philosophy, techniques and ultimately creation. Thus, no exclusive qualification is applicable: it is science and practical reflection, it is technique and art and everything at the same time. The traditional conception places education as an object of treatment of an intermediate science between the positive sciences and the philosophical sciences because it allows the empirical study of the manifestations of the educational process and also the analysis and speculative treatment around the purposes and limits of education, as well as the consequences that derive from it.

León, what is education? University of the Andes. Venezuela; They point out three very important dimensions in every educational process: 1.- Education as a practical task. 2, - Education as a speculative task. 3.- Education as a normative task [12].

Regarding the first consideration of practical task, it is obvious, that to know about education if it does not obtain practical results, nothing solves, and the educational process is not carried out. It is necessary that "art" and that continuous creation facilitating pedagogical interventions appropriate to the multiplicity of situations that occur in the process and for which, the educator as an executor of that art, must possess sufficient imagination, ability to improvisation and as not a good dose of intuition. You can: understand education as art, if art is understood as a set of positive and subjective dispositions to act. Art confers natural disposition.

Regarding the second mentioned aspect, without this being an indicator of order of importance, it refers to the speculative function, and we consider that this aspect must be joined in the process, to the practical aspect. It is not possible to act coherently without prior knowledge of the path we must follow to achieve predetermined goals. This third aspect that refers to the normative dimension is; consequence of the previous one. From this function arise principles that intervene in the process from other areas, but that interact decisively in the educational act: biological principles, principles: psychological, ethical principles.

Pedagogy has an important epistemological content because of its relationship with other sciences whose specificity is the study of man, which are called Human Sciences, and which generate areas of study on which the research is projected; in search of knowledge. Most theorists such as Hoyos [13], Civarolo [14], Barriga [15], Addine [16], Zuluaga [17], Comenius [18], Bruner [19]. They have realize important investigations that become the theoretical referents for this work so many can say that education have been concerned about epistemological studies: About pedagogy and the sciences of education especially in the last ten years, which does not mean that previously there was no such concern for these issues, not only in pedagogy but also in other sciences such as biology, psychology, sociology, etc; sciences that were studied in their etymological aspect by pedagogues, something that was not matched to the same extent by the specialists of those. Such studies were motivated by the deep interrelation that exists between Pedagogy and Biology, Psychology, Sociology, etc.

The epistemological application to the field of the Science of Education is called Pedagogical Epistemology and involves a cognitive investigation in the field: of pedagogical knowledge making

critical assessments in order to determine the logical structure of Pedagogy about the following aspects: a) The research objectives. b) The principles. c) The methods. d); The results.
In the words of Sáez in Epistemology and Pedagogical research it is convenient to keep in mind that the subject and the object form a unit in which they; elements interrelate reciprocally. Pedagogy is an autonomous science of education that has its own character and its purpose is: the study of education understood as an activity of improvement [20].

The pedagogy; As an autonomous scientific discipline, it admits subdivisions that study educational processes from and in other dimensions: Experimental Pedagogy, Didactics, the History of Education, Educational Guidance, etc.

This same author Sáez, suggests that Pedagogical Research and Pedagogy as Science, admits other Auxiliary Sciences, thus avoiding isolationism and radical organicism, and this is how their support base is configured, without implying the loss of identity, but on the contrary, it conforms what Simmel calls educational reality as a whole full of heartfelt [21].

Education as a process and an institution is called more than any other to generate the conditions that allow not only the approach to knowledge, but also the development of an environment that leads to understand and submerge the social and personal need of this knowledge. Myth, religion, philosophy and science are scales of this basic effort to discover realities that directly affect the life of the individual and sociopolitical groups. All of them are modes of reasoning and each one of them individually or by groups have been able, in the course of history to locate basic problems of knowledge and to solve them in their own way or in very special cases like the religious build appearances for solve them. This is why questions like: "What we know" "How we know" have been important points of reference for all those people that through time such as Aguirre, and Jaramillo [22], and Asencio [23]. The different natural phenomena such as earthquakes, lightning, rain and drought, but these events must have natural causes and so the first so-called philosophers of nature arise, to investigate and reflect on how changes in nature are possible, and if there is a single primary substance originating from all the changes. The writings of Hilbert [24], Radice, [25] analyze the history of the Ionian School with Tales, Anaximandro and Anaxagoras. On the other hand, Parmenides says that things do not change, and nothing can be converted into something else, so, for him, the world is illusory and only reason can say it is the world, so the reason is the source of knowledge (Fundamental postulate Rationalist philosophy). Plato mediates between these two positions, by stating that some things change and not after; what changes belongs to the world of the senses or material world and what does not change belongs to the world of ideas or immutable and eternal abstract molds. His student Aristotle, born in Estagira, takes up the subject of what changes and what does not change, but reverses the conception of his teacher considering that the molds are the result of an inductive process on the empirical until reaching the generalization, as well Reality is in the world of the sensible. There is nothing in the intellect that was not first in the senses.

The Aristotelian philosophy dominates the intellectual, philosophical and religious landscape until the Renaissance where the works of Copernicus and Galileo collapsed the scientific conceptions of the Stagirite. With the apogee of the two schools mentioned above, there emerges the figure of the first university professor of philosophy, Emmanuel Kant, who generates a synthesis between empiricism and idealism, "Knowledge of the world comes from perception, but this perception is conditioned by our cognitive structures ", is his primary and fundamental postulate of his thought [26].

With Kant, clearly appears the knowing subject and the object of knowledge, and knowledge is generated through the interaction between these two elements. The subject is born with innate structures sustained in the a priori forms of sensitivity: Space and Time, which allow the subject to grasp an Object, but not as a copy of reality because before it must pass through the sieve of cognitive structures, thus generating the object "for me", because the object "in itself" can never be known by the knowing subject [26].

As a result of the "crisis of the foundations" that affected some of the sciences in the last part of the 19th century, it was demonstrated that the Kantian philosophy could not answer certain questions from science, especially its conception of space was hit hard with the appearance in 1832 and 1854 of the Euclidean geometries. Later Einstein with his theory of relativity would corroborate this collapse.

Reading the following article by Sánchez [27] it can be said that Definitely who produces an epistemological rupture is the intellectual Jean Piaget who, through his rigorous and profound investigations, and this makes him different from the other theorists, formulates a Theory of knowledge: Genetic epistemology. For this Swiss epistemologist, the genesis of knowledge is in "the action" and is the actions on subjects whether physical or mental, which once internalized produce knowledge, but differ from the Kantian position in that there is a rejection of all apriorism and at the same time rejects all empiricist reductionism. So, with Jean Piaget, the subject is an active subject, builder of meanings product of the actions on the objects of knowledge, thus laying the foundations of the constructivist theory of learning.

4. Conclusions

The processes of educational thinking and the possibilities of knowledge in the education sciences are developed using different strategies according to the area of knowledge treated: Natural Sciences, Mathematics, Social Sciences, Spanish Language and Literature.

The study of Logic is the study of the Methods and the Principles used to distinguish the Right Reason from the Incorrect. The distinction between right and wrong reasoning is the central problem that logic must deal with. Although the method of deductive logic penetrates all fields of human knowledge.

In the sixties there is a remarkable development of a series of theoretical methodological currents in the field of Social and Human Sciences that are inscribed within this hermeneutic conception of science which is also known as "Hermeneutic Paradigm", in the one that registers streams like: Ethnography, Interpretive Sociology, Symbolic Interactionism, among others. All these currents postulate a hermeneutic-linguistic phenomenological approach to capture the meaningful, intentional dimension of human action.

There is no knowledge without interest, that is, that all knowledge is governed by interests that give it meaning and that constitute its deep drivers. The Spanish socialist and anarchist emigrants promoted popular education in Latin America since the beginning of the 20th century through union schools, popular universities, cultural movements, among other modalities. The most powerful stage in Latin America took place from the mid-sixties and the eighties.

Investigating social reality is not easy. The researcher is part of the social reality that must be investigated. The subject / object opposition is blurred. For object is what is outside the subject, literally "what has been thrown from the subject", and here the subject is inside the object, and how can we understand what we understand? For subject is what is subject or bound, the object being that of which the subject is loose-that is why it can flee from the object or modify it-and here the subject is bound by the object, imprisoned in the social order that must be investigated.

References

[1] Jarpa A G 2017 *Revista Académica UCMaule* **53(1)** 9-31
[2] De los Rios D and Canales A 2007 Factores explicativos de la deserción universitaria *Revista Calidad en la educación* **26** 173–201
[3] Castaño E, Gallón S, Vásquez J 2008 Análisis de los factores asociados a la deserción estudiantil en la educación superior: un estudio de caso *Revista de educación* **345** 255–280
[4] Fernández M 2008 Hacia una pedagogía de las diferencias desde los aportes de la propuesta de Paulo Freire. Ed. Moacir M, Gomez V, Madra J and Fernndez de Alencar 2008 A *Paulo Freire. Contribuciones para la pedagogía,* (Buenos Aires: Consejo Latinoamericano de Ciencias Sociales) pp. 341–348
[5] Acuña C 2012 *Accesso y Deserción en la Educación Superior, Caso aplicado a Chile* (Santiago de Chile: Universidad de Chile)
[6] Bolívar A 2005 Equidad educativa y teorías de la justicia *Revista Electrónica Iberoamericana Sobre Calidad, Eficacia y Cambio en Educación* **3(2)** 42–69
[7] Olave-Arias G, Rojas-García I, Cisneros-Estupiñán M 2014 Deserción universitaria y alfabetización académica *Revista Educación y educadores* **16(3)** 455–471
[8] Becerra-Restrepo M M 2017 *Pensamiento Pedagógico Colombiano Una mirada a sus conceptos de pedagogía, infancia, maestro y escuela* (Bogotá: Editorial Magisterio)

[9] Baudrillard J 2012 *Cultura y Simulacro* (Barcelona: Kairós)
[10] Herrera J D 2013 *Pensar la educación, hacer investigación* (Bogotá: Universidad de la Salle)
[11] Bolívar A 2016 Educar democráticamente para una ciudadanía activa *Revista Internacional de Educación para la Justicia Social* **1(5)** 69–87
[12] León A 2007*¿Qué es la Educación?* (Venezuela: Universidad de los Andes)
[13] Hoyos C A 2010 *Epistemología y objeto pedagógico ¿Es la pedagogía una ciencia?* (México: UNAM, Instituto de investigaciones sobre la Universidad y la Educación)
[14] Civarolo M M 2011 *La idea de la Didáctica* (Bogotá: Magisterio)
[15] Díaz-Barriga A 2012 *Pensar la didáctica* (Buenos Aires: Amorrortu)
[16] Fátima A 1998 *Didáctica y optimización del proceso de enseñanza-aprendizaje* (La Habana: IPLAC)
[17] Zuluaga O L 1999 El saber pedagógico: Experiencias y conceptualizaciones *Encuentros pedagógicos transculturales: Desarrollo comparado de las conceptualizaciones y experiencias pedagógicas en Colombia y Alemania* (Medellín: Universidad de Antioquia) pp 81-88
[18] Amos-Comenio J 2012 *Didáctica Magna* (Madrid: Akal)
[19] Bruner J 2012 *La importancia de la educación* (Barcelona: Paidós)
[20] Sáez A R 2017 La prioridad del método en la investigación pedagógica *Revista Española de Pedagogía* **75 (267)** 239-254
[21] Simmel G 2008 *Pedagogía escolar* (Barcelona: Gedisa)
[22] Aguirre J C and Jaramillo L G 2010 La ciencia y el sentido común: por la enseñanza de un sentido común crítico *Educación y Educadores* **13(3)** 477-494
[23] Asencio Cabot E 2014 Una aproximación a la concepción de ciencia en la contemporaneidad desde la perspectiva de la educación científica *Ciência & Educacao* **20(3)** 549-560
[24] Hilbert D 1950 *The Foundations of Geometry* (Illinois: The Open Court Publishing Company) pp 2–16
[25] Lombardo R L 1983 *La matemática de Pitágoras a Newton* (Barcelona: Editorial Laia) pp15–17
[26] Mansur J C 2009 Principios precríticos y críticos del pensamiento de Emmanuel Kant *Revista de Filosofía* **36** 67-84
[27] Sánchez M A 2016 El proceso de construcción de normas sistémicas visto desde una epistemología genética neoconstructivista basada en Jean Piaget *Papel Político* **21(1)** 123-166

IOP Conf. Series: Journal of Physics: Conf. Series **1161** (2019) 012006 doi:10.1088/1742-6596/1161/1/012006

Digital inclusion: Strategy of reconstruction of the social fabric

D A Prada[1,3], A Acevedo[2], P E Vera[1], F A Páez[3] and J M Gómez[4]

[1] Grupo de Investigación en Materiales GIM, Universidad Pontificia Bolivariana, Bucaramanga, Colombia

[2] Grupo de Investigación en Administración GIA, Universidad Pontificia Bolivariana, Bucaramanga, Colombia

[3] Grupo de Investigación SED Universidad Pontificia Bolivariana, Bucaramanga, Colombia

[4] Centro de Gestión de Mercados, Logística y Tecnología de la Información, Servicio Nacional de Aprendizaje SENA, Bogotá D.C., Colombia

E-mail: duwamg.prada@upb.edu.co

Abstract. Digital inclusion is seen as one of the strategies that allows people linked to different social, academic, labor groups, among others, to develop certain activities without having to make physical presence in a specific meeting place. However, talk of inclusion is not limited to the simple acceptance, demand a compromise which covers not only share knowledge, but rather generate a tool that tends to learn and learning to learn. Identify the possible factors that affect ownership and evidence that computer literacy is not delimited in a type of mechanistic approach but must also be supported in a continuous construction of knowledge are it is the central idea of this writing. Also, it specifies that Information and Communication Technology can't be converted in differentiating agent, or that it is designed to measure a community that identifies its citizens skill in the use of computer resources, but Information and Communication Technology should serve as a tool to the continuing evolution of society without leaving aside its history, custom and culture.

1. Introduction

With the continuous increase and ease of acquisition of technological tools, interaction and the need for learning in the use of technological devices, it has become an everyday activity for any community. Recent studies show that 80% of Colombians use the Internet and that the socio-economic strata that most use it is stratum 1 and 2. The possibility of digital inclusion in terms of Internet coverage is a reality, however it is evident the need to optimize such inclusion in various fields, for example the social one. Generate digital inclusion in Colombia, will allow the solid construction of processes of equivalence of opportunities in economic terms to promote the reconstruction of the social fabric, deteriorated in various areas because of the digital divide.

2. Inclusion

The term Inclusion is perceived as the total integration of a group of people with respect to a particular topic, that is, to make participation part of inclusion allows us to live in common unity. Currently and for several years, attention to diversity in a digital context constitutes one of the fundamental and priority challenges at the international level [1]. Digital inclusion has become one of the strategies of greatest need and need in a society that continually evolves in favor of the optimization of computing resources, in addition, seeks to effectively reduce procedures that generate unnecessary time investment for a user of a respective activity. In an economic activity, it is very useful to make transfers, payments or purchases by electronic means, since the expenditure of money and travel time is avoided. Regarding the activity academic, there are books, texts and articles in digital form, which allow investing in other activities, the time that is used in the bibliographical consultation. Electronic resources are increasingly used because they have become a tool that minimizes the time in execution of a particular task.

2.1. Digital literacy

With the technological advancement, both in hardware and software, the user can develop different tasks in a more efficient way, however, this requires a progressive improvement in the skill with respect to the use of the resource, that is, to be updated and trained. Continuously due to the interaction with the tools. In the last decades, digital literacy programs have become a governmental priority through which it is intended, on the one hand, to contribute to the integral development of the country and, on the other, to reduce the socioeconomic gaps of its population [2]. In relation to the above, the need to create mechanisms in which people can access such digital literacy is notorious, however, it should be clarified that in addition to this it is essential to generate a culture in which users continue independently their own learning in front of the new technological tools that are in the middle, in training the person trained integrally is autonomous in their learning, is able to generate their own criteria based on in the things they observes, looks, listens and reads[3]. Then it is possible that the question becomes evident: are universal access and computer literacy the keys to ending the digital divide? Socioeconomic inequality and social exclusion require a more aggressive and nuanced plan to deal with this problem and cannot be based solely on the transmission of information or only on the teaching of competencies [4], because you must understand the learning as building and restructure models to interpret the information it receives [5]. They advocate a digital human capital framework, taking into account the complex nature of social exclusion in the information age [6].

It is possible to observe that the answer to the previously formulated question is very subjective, because it is a punctual question and thinking about computer literacy contemplates a series of questions that do not depend exclusively on their cognitive abilities. Computer literacy is not just a series of rules to follow to learn how to use a technological tool, such as the use of a computer or the use of a specific application or software; This type of learning or knowledge is routine and sequential, knowledge from a mechanistic approach conceives the subject as a passive being who only receives information from their environment [7].

The information and knowledge society demand a growing digital literacy of its citizens. Education, in formal and informal settings, has a remarkable role in favoring inclusion and social inclusion, since it helps to develop skills that allow access, record, edit, publish and share content on the network, autonomously, critically and responsible. In this regard, it should be noted that governments must provide the necessary resources to make this possible [8].

3. Academic strategy for inclusion

For the development of an optimal strategy for inclusion, it is essential to identify which are the most obvious factors that influence the use of technological tools. The key factors, with their corresponding proposals for action, as essential to favor the changes that must occur in education are strategic management, generalization of access to technology, ongoing teacher training and evaluation and monitoring of policies and actions that promote digital inclusion [8]. This type of factors contains several elements that seem imperceptible, one of them as a factor that directly affects digital inclusion is that of resources, since it is necessary to have the tool to carry out processes in which it is possible to practice what has been learned and enter, this is where the idea of starting with the technique comes from. The computer is used as a support tool in academic, economic, communication processes, the elaboration of personal and social files, in simulations of various models of natural phenomena, etc. Another factor that influences is knowledge, because as companies develop more technological devices, they bring with them new tools, new applications that generate ease of use, but when they are ignored, we underutilize these tools. For this reason, it is very important to generate the culture of self-learning, that is, not only learn but learn to learn [9].

Taking into account that the information and communication technology (ICT) tools are a source of support and that they are in constant and continuous change; It requires a learning methodology that allows acquiring the ability not only to learn to use a certain computer tool, but also to strengthen knowledge of it, and generate self-learning of any other tool. In the search for the methodology, the Universidad Industrial de Santander designed and developed an educational process to learn how to use computer tools, which was called "self-learning system for the use of computer tools" (SAHI) whose general objective is learn to learn and learn [9]. Even if you have the necessary tools and competence in the use of certain tools, it is necessary that quality content be provided digitally for the continuous improvement of the knowledge society, and we cannot ignore that The Internet is a tool that allows better and greater access to information, under a more efficient form of communication, for this reason it is important to analyze the efforts of computer scientists to ensure the authenticity of online comments that are crucial for e-commerce and social media sites [10,11,12].

Social inclusion is seen as the search for the reduction of inequality, a balance between the rights and duties of individuals and the increase in social cohesion [13]. Digital government has become a catalyst to allow social inclusion through greater access to services and the democratic process. However, it is essential to ensure that those who lack access, skills and / or the desire to make use of technology are not excluded from access to information and / or the democratic process [13].

4. Results and discussion

In Colombia, the ministry of information and communication technology (ICT) and Ipsos Napoleón Franco carried out and presented a study on digital consumption in March 2014, which showed that 8 out of 10 Colombians are using the internet and the greatest increase in the use of the network occurred in the strata. 1 and 2. It was also observed that 54% of Colombians who use the internet use it daily and invest between 2 and 3 hours browsing. 64% of the houses in cities with more than 200 thousand inhabitants have a connection, 71% of the respondents access from their home and 20% from an internet cafe. 26% connect to perform free courses such as those offered by Servicio Nacional de Aprendizaje SENA, 24% do it to find employment and 22% to quote products. Regarding the administration of their free time it was found that 30% of respondents watch movies from a computer, 15% watch movies online and 12% download them, 40% listen to radio from their cell phones. Finally, the survey that was conducted was able to establish how digital Colombians are, 33% are in the group of novices and interested, 31% in digital advanced, 19% are unaware, 12% are the most curious explorers and 6% are the apathetic internet [14].

To demonstrate the dynamics of using digital tools with respect to the instruments that are acquired such as the computer or the Smartphone, a measurement instrument was applied to 80 people each in a small or medium-sized company in the city of Floridablanca in the department of Santander Colombia. This sample size was taken from a population of 264 companies with a 90% confidence level of 1.28 and a possible measurement error of 6%. The instrument investigates sociodemographic data and then addresses knowledge and the use of technological tools, then questions are asked about the applications that most use. The instrument was applied to 67 women and 13 men, 27 of the 80 respondents are between 18 and 27 years old, which is equivalent to 33.8% and only 13 people surveyed are older than 47 years, which is equivalent to 16.3%, where 43.8% live in a stratum 3. Regarding the use of smartphones, 80% said they had one and of these 25.8% said they had a data plan with Internet payment, that is, the rest uses the Smartphone through networks wireless wifi at home or in the workplace. Regarding the question about the application that 67 people used the most, corresponding to 83.75%, they answered that WhatsApp.

With the above, there is an allusion to what we know today as the technological gap, that is, although we recognize the valuable contribution that advances in computer resources have, there are obstacles that keep us away from interaction with these tools and we focus on the use of them in basic applications that do not need too much physical resources. The sectorial report of the ministry of information technology and communications shows the access gaps that persist in the interior of the country. In the first quarter 2013 report, a total of 6634659 internet connections were recorded, of which 4013553 are fixed and 2621106 mobile. The highest penetration rates per subscriber (over the total population) are led by Bogotá with 16.45%, Antioquia with 12.54%, Risaralda with 11.63%, Santander with 11.17%, and Valle 9.9% The penetration in departments like Córdoba, Cauca, Nariño, Arauca, Chocó, Caquetá, La Guajira and Putumayo does not exceed 3%, and in the case of the national territories Amazonas, Vichada, Vaupés, Guainía and Guaviare does not reach 1% [15].

Finding possible reasons why people who are part of this computer illiteracy do not want to change to acquire skills in the use of technological tools is a difficult task, as well as assessing the effective impact of each of the components of the digital ecosystem in technological diffusion where the persistence of regional asymmetries warns of the urgency of applying more equitable criteria of digital inclusion [16].

As a consequence, the questions arise: what and how should a person be taught to acquire skill in the use of electronic devices? Although the question does not present an immediate answer because as human beings we change continuously and also that we learn in a different way and time, it is difficult to find the answer to these questions, however, if there is a methodology that works with some people, it is not possible to guarantee that it will serve everyone. In Colombia, a strategy was implemented to provide computers to official institutions in any region of Colombian territory. This strategy was an implementation of an action carried out in Canada known as computers for schools, an implementation known by the first lady of the time in which she observed the benefits of this idea and proposed to develop it in Colombia under the name of computers to educate. At the request of the president, the national planning department prepared the compilation document with the number of file 3063 for the national council of economic and social policy, which approved the Program and entrusted its development to the ministry of communications, ministry of education and SENA [16].

The program begins on March 15, 2001 with the collection of computer resources, basically computers that have been written off by public or private companies. These computers are re-powered and delivered to the different selected institutions, which mostly had access to the internet through the compartel program. One of the factors of vital importance for the economic, social and cultural development of Colombia is the use of the opportunities that the information and communications technologies have to offer. The leaders of the public and private sector, educators and parents share the belief that the future of Colombia's economic growth will be based on the effectiveness of our innovation, adaptation to technological change, and taking advantage of opportunities to create jobs, promote equity social and generate well-being [16].

However, given the proposal to provide institutions in the rural sector with computers and the internet, it is clear that the need to learn how to use the tool is not met, much less the digital divide is reduced. It is necessary that an accompaniment be carried out in the maintenance part of the resource, and regarding maintenance it is not only the operation but also that it starts with activities to improve the right place, that is, that activities begin to acquire new equipment and also an educational accompaniment so that activities are carried out so that the people of the community can interact with the tool.

The adequate application of information and communication technologies facilitates having: a more efficient, transparent public sector with better services for the community, a more competitive private sector with greater possibilities of entering the international market and a better prepared community to face the challenges of the modern world, with greater mechanisms of participation and communication, and with unlimited access to information and knowledge [16].

The search for social equity based on the idea of inclusion allows the other to be recognized as a fundamental part of the environment to which it belongs. Under this premise, and with the continuous development of ICT in the government, the concept of digital government has been built, whose benefits are: to facilitate more convenient government services for citizens and businesses; improve economic development; remodeling and redefinition of the community and government processes; what allows greater public access to information and make governments more accountable to their citizens [13].

With the benefit of greater access, greater field of action on the part of the government entity and through the use of the internet, it is possible to know firsthand the different problems of a community, such as the birth and death rate of a municipality, the access roads to this place, the number of inhabitants, the climate, customs, cultural activities, its history, among other items that characterize a community. This type of activity in which it is possible to know more about one's own culture and the roots of it, generates the space in which people recognize themselves as important active agents in the development of their environment. To be digitally literate is to possess the technological training necessary to survive in the Information Society and to act critically on it. The digital literate society is that composed of active subjects that intervene decisively in matters that affect them, use technologies with competence and critical capacity, and do not resign themselves to the role of mere consumers of technologies and digital content. Digital literacy is important because it is the key to the inclusion and development of an Information and Knowledge Society for all people [17].

5. Conclusions

The ICT should not be the purpose or scale of measurement of our development but the tool in which we can support everything we are, what identifies us as citizens of a digital world and in continuous evolution, without leaving aside all that that has identified us as beings with a history, a culture and builders of a community.

It was possible to observe that from a population of 264 employees of 80 the micro, small and medium enterprises or companies, an instrument was applied with 12 questions about the use of technological tools, in which it was evident that 80% of the respondents had a Smartphone and considered it a tool important as support for their work, basing the above only on the use of email and WhatsApp. Digital inclusion is not only focused on the mechanistic learning of a particular tool, because at every moment we observe in the download stores, various applications that can facilitate different processes, but due to the fear of facing these new programs, people they only keep what they have learned, that is, they did not learn to learn. The ICT cannot become the differentiating agent, nor that its purpose is the measurement of a community that identifies its citizens by the skill in the use of computer resources, on the contrary they should serve as a tool that allows the continuous evolution of the society without forgetting its history, its custom or its culture.

References

[1] Cabero J and Córdoba M 2009 *Rev. Educación inclusive* **2(1)** 61-77.
[2] Román E 2009 *Cal. Edu.* **30** 187-205
[3] Martinez F 2009 *Rev. Doc. Uni.* **10(1)** 123-136.
[4] Restrepo L 1996 *Proyecto de educación sexual* (Bogotá: Ministerio de Educación Nacional)
[5] Perez M 2011 *Principios de aprendizaje* (Bucaramanga: Universidad Industrial de Santander)
[6] Bach A, Shaffer Q and Wolfson T 2013 *J. Inf. Pol.* **3** 247-266
[7] Acodesi 2002 *La formación integral y sus dimensiones* (Bogotá: Kimpres)
[8] Lázaro J, Estebanell M, and Tedesco J 2015 *Rev. Uni. Y Soc. Con.* **12(2)** 44-58
[9] Andrade H and Gómez L 1997 *Tecnología informática en la escuela* (Colombia: Universidad Industrial de Santander)
[10] Huaroto C 2012 *Inf. Tech. & Inter. Dev.* **8(4)** 129-144
[11] Kugler L 2014 Keeping Online Reviews Honest. *Com. ACM.* **57(11)** 20-23
[12] Martínez J and Torrent J 2014 *El Profesional De La Información* **23(3)** 288-299
[13] Asgarkhani M 2007 *J. Tech. Hum. Serv.* **25(1-2)** 127-146
[14] Ministerio de Tecnologías de la Información y las Comunicaciones Min TIC 2015 *Boletín Sectorial de las TIC* (Bogotá: Ministerio de Tecnologías de la Información y las Comunicaciones)
[15] Ministerio de Tecnologías de la Información y las Comunicaciones Min TIC 2013 *Boletín Sectorial de las TIC* (Bogotá: Ministerio de Tecnologías de la Información y las Comunicaciones)
[16] Computadores para educar 2016 Consulted on: http://www.computadoresparaeducar.gov.co/PaginaWeb/index.php/es/nosotros-2/historia
[17] González N 2015 *Qué puede hacer la biblioteca 2.0 por la inclusión digital* Consulted on: http://nieves.pbworks.com/f/Biblioteca2+e+inclusion+digital.pdf

IOP Conf. Series: Journal of Physics: Conf. Series **1161** (2019) 012007 doi:10.1088/1742-6596/1161/1/012007

The principles of sports training as a methodology alternative in the cognitive development of the human being

D A Vera-Rivera[1, 2], N P Guzman-Pinzon[1] and D F Rodriguez-Neira[1]
[1] Universidad Pedagógica Nacional, Bogotá, Colombia
[2] Fundación Universitaria Área Andina, Bogotá, Colombia

E-mail: daverar@pedagogica.edu.co, npguzmanp@cpedagogica.edu.co

Abstract. This research focuses on establishing a relationship between the principles of sports training and the processes of cognitive development of human beings whose general objective lies in the design of a methodological alternative. In this sense, sports training are understood as one of the most widely disseminated and researched topics on an international level in the field of sports, where there is a wealth of theories and trends that help to base and investigate human development processes; hence it is understood as one of the central categories of this work. In this regard, Bautista [1] calls us to recognize sports training as a pedagogical model in the educational field. Therefore, it is very important to establish as fundamental parameters the classification of training principles, divided into two large blocks by Muñoz and Soto [2], the Biological principles and the pedagogical principles. In this thread, it is important to highlight the category where the above is based, cognitive development understood as the axis that implies the identification of changes that influence the different representation structures that operate in the cognitive system and that point to the rediscovery and reconstruction of new knowledge Hence, the guiding question raises: What is the relationship between the principles of sports training and the processes of cognitive development in the human being? Thus, the methodological route of the present study responds to the interpretative paradigm, specifically it is oriented from the qualitative design. Therefore, it exposes an approach to the establishment of possible relationships between the categories of analysis.

1. Introduction

"Neither the athlete nor the coach (teacher) are producers of passenger sports results, but mediators of intellectual skills (intelligence actions), adaptive skills transformers owned by those who learn them" [1].

Throughout time from the field of psychology has tried to identify, analyze and understand How does the human being learn? As a result of this, strong theoretical conceptions have been known for many years to explain this process. Among the most relevant are the Piaget [3] and Vygotsky [4] approaches, and the epistemological orientations of Ausubel do not depart from them [5]. They have been theoretical conceptions that guide the teaching-learning processes in schools, including the forms of relationship of the subject in society. On the other hand, the training itself, from a rudimentary perspective or if you want to read from the traditional, was conceived as the opportunity of man in terms of preparing for life in terms of survival. The new theoretical and practical constructions of this field indicate that the objective is not only the preparation of the high-performance athlete, but the stimulation in the motor state focusing on the cognitive, motor, affective, axiological, social contents proclaimed as well as the science that seeks the training of the athlete in a whole [6]. Based on the above, the question that leads

IOP Conf. Series: Journal of Physics: Conf. Series **1161** (2019) 012007 doi:10.1088/1742-6596/1161/1/012007

this proposal raises: How to establish a methodological alternative from the principles of sports training, aimed at the processes of cognitive development in the subject? Specifically, the bibliographic tracking does not account for investigative processes that pose different relationships between the field of sports training and cognitive development. However, it is worth highlighting research that recognizes sports training as a mediating axis of socialization processes, behaviors and behaviors through the preparation of the athlete in their different ages and contexts.

In this regard, Bautista1 highlights "sports training as a pedagogical practice that is interdependently related to the phenomena and processes that allow the development of the individual within the scope of his knowledge, that is, his intelligence". In this line, Carrasco [7] affirms that the athlete participates actively in training, knowing what he does, how he does it and what he does. At the same time, it contemplates that the preparation and conduction of the training must be oriented towards a task that allows the trainer and his trainees to know the reason and why of the task [7]. In addition to the above, Levpavlovich states that "The theory of sports includes the theory of sports training supported by the theory of development and the theory of education of the individual" development and education "in this sense is understood the process of the transformation of their properties during the stages, periods and cycles quite prolonged of the vital time of the subject" [8].

According to the above, to refer to sports training involves recognizing the aspects that make it up and nourish it, within these aspects as a fundamental pillar are the principles of sports training classified in pedagogical and biological as referents from the methodological and adaptive, defined according to Carrasco [7] as the guidelines that are strengthened in the different sciences: pedagogical, psychological and biological, enabling the trainer to adequately implement global content, methods and planning, as well as control over the trajectory of preparation. For this case, the hypothesis of this study lies in conceiving the principles of sports training as a methodological alternative in the process of cognitive development of the subject. It should be noted that the above responds to the first phase of the construction of the investigative process, in this sense, it gives an account of the conceptualization and contextualization, as opposed to the first relational approach between the main categories of analysis: principles of training (pedagogical, biological and psychological) and cognitive development.

2. Methods

The present study is based on the interpretative paradigm. Specifically responds to the qualitative design, because at first (conceptualization phase) required review processes clearly analytical and interpretive in order to build possible conceptual and practical relationships between the main categories of analysis. With the purpose of establishing (cross-contextualization phase) transversal relationships between the aforementioned categories of analysis, and thus envisioning a methodological strategy that poses emerging dynamics in the teaching-learning process.

3. Results and discussion

The present conceptual approach in the first instance refers from the theory to the subject of sports training from the principles, those expressed from physiology as theoretical elements that try to explain the organic function of the subject, and those recognized from the pedagogical field, referred to the cognitive and axiological dimensions of it. In this sense, it is necessary to affirm that sports training from a biological perspective is defined as "the systematic repetition of muscular contractions above its threshold producing morphological and functional adaptations" [9]. From the pedagogical perspective, it is recognized as the orientation of the structure and organization of the configuration of the human being. In this way, it is clear that the concept allows the teacher and the student to be placed on the scene as protagonists of the development of knowledge, who are irreplaceable within the teaching-learning process, and who are in charge of recognizing and understanding the complexity of their interaction.

In general, the intention of sports training should start from the conception of the whole, not the particular, so that the subject is valued as an integral being that manifests and shapes from the following relational spheres: affective, volitional, cognitive and psychomotor, making it a thinking, sensitive and in need of movement. The foregoing allows to state that sports training is assumed as a methodological

IOP Conf. Series: Journal of Physics: Conf. Series **1161** (2019) 012007 doi:10.1088/1742-6596/1161/1/012007

process and adaptation, which in terms of the cognitive, can be assumed as a possible methodological mediation that enables the modifiability of the cognitive structures of the subject who learns. To focus more specifically on the category object of interpretation, the following are related from Solano and Nuñez [10] the principles of training considered as fundamental (see Figure 1).

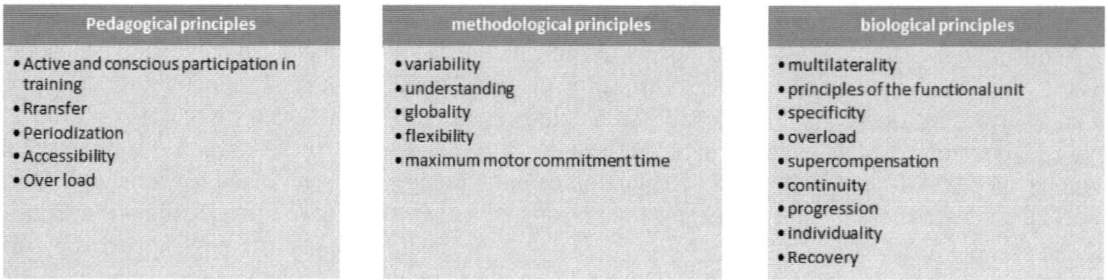

Figure 1. Principles of training [10].

The above, allows to identify the requirement to which the subject is subjected from its organic functions, from where it begins to raise changes not only in terms of the development of physical abilities, but also affective-emotional elements that influence sports training.

In this sense, the importance of the balance / stability necessary between the affective-emotional factor and the physical-biological is recognized, so maintaining a clear and coherent intentionality will allow establishing successful sports adaptation processes in terms of the development of the subject. Therefore, it is valid to project the structural changes according to a subject that through sports practice can carry out social actions that allow him to bring the partner or a group-team closer to achieving their sporting goals, whether individual or common, and perhaps overcome the adversities of the context from the development of physical, technical, tactical and psychological capabilities. Therefore, pedagogy as a field of reflection begins to gain value in the arena of sports training, because any process with a pedagogical perspective allows us to recognize the different appreciations and relationships that make possible the configurations of new conditions and possibilities of doing and being.

From the disciplinary approach is important to address the views that are held from the psychology on the concept of pedagogy, in this regard is recognized as the evolutionary processes in educational environments that allow to establish solutions to a problem, for example "the pedagogy is not another which is the extension of modern methods of psychology to properly pedagogical problems, and especially to school problems "[9, 10]

Another recognition of the term is seen from the creative opportunity, while establishing the pedagogy as a field of recontextualization of sports training, and against this, the deconstruction of the priority given to the expression of the result, establishing a new priority riches of social interaction where several elements converge that can strengthen a better process of the cognitive development of the subject starting with sports training.

Thus, the analysis discussed so far addresses the training as a pedagogical, biological and methodological process taking into account that the literature leads us to the search to achieve a goal determined by the planning of the loads, without ignoring the appropriate recovery processes, providing stimuli individually, progressively and multivariate.

Regarding the category of cognitive development, it is feasible to recognize from the different interpretations that respond to how does the subject learn? It is not a matter of simple biological nature but promulgates by a multidisciplinary work that implies to recognize in the first instance that it is a process that includes traits of biological, psychological and socio-cultural character. To start the discussion, we start from the postulates proposed by Piaget, who argues that the construction of knowledge of the human being given in an educational and formative context such as school is complemented by the intellectual and physical development in which it is mediated. Broadly speaking, the theory put forward by Piaget [3] refers to six stages of development:

1. Stage of reflexes or hereditary adjustments.
2. Stage of the first motor habits and the first organized perceptions.
3. Stage of sensory-motor intelligence.
4. Stage of intuitive intelligence.
5. Stage of concrete intellectual operations (start of logic).
6. Stage of abstract intellectual operations.

The previous stages are recognized by Piaget as structures of variable character that start from motor/intellectual aspects, individual and collective, in addition to the affective, that is to say that it is not a finished process, on the contrary it is unstable, changing and dependent [3]. For his part, refers that learning can be developed from the inductive (discovery) and deductive (reasoning), in turn says that the "individual" reaches an optimal development of learning when he starts contact with the object of study from his experience. In this regard, it points out three stages:

1. Symbolic (makes use of a concept and approaches abstract thinking).
2. Iconic (makes use of images and approaches imaginative thinking - projective).
3. Enactive (makes use of actions and intuitions and approaches a concrete thought, therefore manipulee - moldable).

As a complement to the above, Ausubel [5] points out that learning in the human being occurs when it transcends the memorization of content, when it manages to find meaning and meaning in what it learns. That is, when new ideas can be learned and retained, and their organization is clear and available to be anchored to existing concepts. This learning process signals itself processes of compression, transformation, storage and use of information. Ausubel [5] manages to distinguish three types of learning:

1. Learning of representations: the human being attributes meaning to symbols by associating them with their objective references.
2. Learning concepts: the human being abstracts from the objective reality those common attributes.
3. Learning of propositions: ideas that result from a logical combination between terms and symbols.

As a complement, Vygotsky [4] develops his theory from the socio-cultural perspective, from where he defines that learning is a social and cultural - collaborative construction, so that human beings can enhance their development from interactions and interrelations with other Humans. Consequently, the development of the human being must be understood as a socially constructed goal.

Therefore, to speak of cognitive development implies to recognize the development as a theoretical-practical category that in a first moment reconfigures the discourse of the mediator-teacher-teacher-teacher, because it calls it to question itself for what happens to the subject, for its transformations, configurations and their mediations. Second, it involves testing what is considered relevant for the subject to learn, in this line thinking the ways as the subject learns and deconstruct and reconstruct and contextualize (without ignoring) the theories and postulates, depending on the intentions of the subject educational act.

In this sense, Antoranz and Villalba define development as "the changes that the human being lives throughout its existence" [12]. To this are added determinant features of development as inheritance, understood as the set of features and characteristics that transmit parents, the environment as the set of circumstances surrounding a person, maturation as the differentiation of biological structures and functional capacities, and learning as "the acquisition of knowledge, skills and abilities [12].

So, the development derived between the feature of maturation and learning (ie the cognitive) assumes a number of characteristics, from Antoranz and Villalba should be progressive, continuous, cumulative, directional, differentiated, organized, holistic, stable/changing, variable, orderly, cyclical and repetitive, finally as a fundamental feature must recognize the individual and cultural differences of each subject [12].

By way of closure, it is pointed out that it is worth establishing relationships between what is proposed from the principles of training and approaches to cognitive development. Well, the educational act requires a general structuring of a plan of action that does not differ from the conditions mentioned when starting the above, namely the biological, psychological and socio-cultural. As a result of the foregoing, a possible transversal relationship is established between the categories of analysis, based on a methodological strategy that enables cognitive changes in the subject based on the reconfiguration of the contents (see Figure 2).

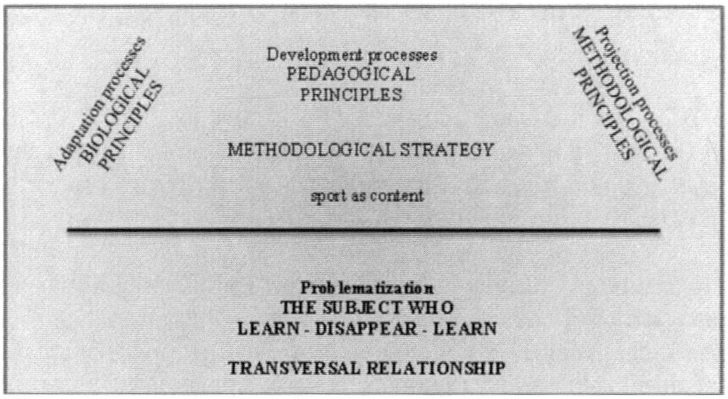

Figure 2. Transverse relationship and methodological strategy.

4. Conclusions

By way of closure, we note that the educational act requires a general structuring of an action plan that does not differ from the conditions and possibilities of the subject who learns namely the biological and socio-cultural.

The principles of sports training are considered methodological horizons that could enable changes in traditional structures that are established from the academic in terms of processes of cognitive development in the subject.

All process of construction of knowledge, in addition to answering a guiding question, is committed to convene and motivate the scientific community to keep generating questions, in this sense what calls to continue building academy, involves testing what is considered pertinent that the human being learn, in that line deconstruct and reconstruct the theories and postulates that are in function of the intentions of the educational act.

References

[1] Bautista Jimenez R 2001 El entrenamiento deportivo como modelo pedagogico de construccion *Lúdica Pedagógica* **1(6)** 37-47
[2] Hernandez Rodriguez A I, Martinez Muñoz L F and Aguila Soto C 2008 *El deporte en la sociedad contemporánea* (Almería: Editorial Universidad de Almería)
[3] Piaget J 1991Six studies of psychology (Barcelona: Editorial Labor S.A.)
[4] Vygotsky L 1979 *The development of higher psychological processes* (Buenos Aires: Editorial Grijaldo)
[5] Ausubel D 1976 *Educational psychology. A cognitive point of view* (Mexico: Editorial Trillas)
[6] Kuhlow C 1975 *The science of training as a scientific discipline* (Madrid, Editorial Paidotribo)
[7] Carrasco F 2006 Training and skills *Scientific Journal* **2** 23-26
[8] Levpavlouch L 2001 *Theory and methodology of sports competition* (Spain: Editorial Paidotribo)
[9] Hohman A, Lames M, Letzelter M 2005 *Introduction to the science of training* (Spain: Editorial Paidotribo)
[10] Solano C and Núñez D 2011 Training adaptation processes *Scientific Journal Health Sciences* **6** 33-36
[11] Gaston M 2010 *Psychology of Education* (Mexico: Editorial XXI Century)
[12] Antoranz E and Villalba J 2010 *Cognitive and motor development* (Madrid: Editorial Editex)

IOP Conf. Series: Journal of Physics: Conf. Series **1161** (2019) 012008 doi:10.1088/1742-6596/1161/1/012008

Soccer teaching methods in sports training schools in Tunja, Boyacá

R D Merchan-Osorio[1], J Florez-Florez[1] and W F Caro-Cely[1]
[1] Grupo LERD, Fundación Universitaria Juan de Castellanos, Tunja, Colombia

E-mail: rdmerchan@jdc.edu.co

Abstract. This study comes from the academic activities carried out by the students of the subjects of Introduction to Sports I and II of the academic degree program in Physical Education, Recreation and Sports of the Fundación Universitaria Juan de Castellanos, Tunja, Boyacá. The purpose of this study was to describe the methods of soccer education that are most used by coaches to train children in the Sports Training Schools of the city of Tunja, Boyacá during the training sessions held along the week. The study involved 20 coaches who belong to sports clubs and sports training schools in soccer categories for children and youth. A methodology with a descriptive quantitative approach was implemented. The information collected was through a semi-structured interview taking as reference a series of questions that were recorded in a file created for such purposes. The data collected were processed using the Microsoft Excel 2016 program for its respective analysis. The results show that the coaches use and apply more, the method of mixed education, which is the combination of analytical and global methods to impart the sports content of grassroots and competitive soccer.

1. Introduction

Soccer is considered one of the most popular sports in the world. The purpose of this is to obtain greater goals than the opponent, is categorized in the cooperation-opposition sports. It can be considered that in this sport the most favorable moment for the child to begin the practice of this sport discipline is in the ages of 6 years and older, since the teaching-learning process depends on the strengthening of the basic motor skills and thus the apprentice can go knowing the fundamental rules and principles of the game that allow him to have a greater degree of independence to solve the problems of the discipline.

In soccer, the child and the young can learn and practice on their own, but for that learning to be correct, the child needs a teacher or coach who knows the process that he or she must follow to assimilate their characteristics, their interests depending on their age, more than that they must know about the sport, the technical and tactical basis, the coordinative development that soccer demands and above all the methodology that the coach must use, so that all these aspects are taken in advance by the child.

According to [1] "In the process of the teaching/learning of soccer, it should be tried to use effective teaching methods, that is, those that make it possible to assimilate learning more and better"

At present, there are different opinions on what the best way to learn a sport, especially soccer, can be. In response to this lack of consensus, several works of research have been developed that aim to demonstrate which approach is the most effective.

In the literature we can find studies that compare the effect of different models and teaching methods on soccer learning that frame it in main aspects such as "the capacity of execution (technique), the capacity of decision (tactical), the capacity of Interaction (Psychological)" [2-7]. The lack of unanimity in the used designs gives controversial results [8] "And limits the possibility of solving some of the most important questions related to learning sports".

According to [9] "defines the teaching method as the way to present the student with activities for their learning that can be global, analytical or mixed, although the specialized literature addresses the systemic method and others that are used empirically and that have no relation to those raised by this author. In this way, it allows us to have a more accurate and concrete vision of the conceptualization and aims of each method of teaching that is explained below: "Global or integral method: The player is faced with a problem situation and decides how to solve it, choosing the action that he believes most convenient, instead of using, in a compulsory and directed way, the one determined by the coach. Therefore, the overall method simultaneously affects technical, tactical, and physical aspects, which allows, during a game, to develop them all together, even if it is presented in a generic way" [10].

Analytical Method: In this method, unlike the previous one, the player performs a series of activities specifically aimed at learning, from one or several basics in an isolated way to the game. In this teaching approach the objective of each activity or exercise is clearly defined, therefore, the player deduces consciously and executes each of the movements necessary to reach the technical task entrusted to him. In this method, it is difficult to keep the motivation and performance of the players for long periods of time, since the proposed situations are not very stimulating because they do not have the playful element, which is continually given in the game [10].

Mixed Method: It is about combining both methods, global and analytical, trying to get the positive performance out of each one. It is the most used method by coaches. The development of this method starts with global tasks followed by analytics and ending with global activities.

In relation to the above, the study aimed to describe the methods of teaching soccer that is most used by coaches to train athletes in children and youth categories in the Sports Training Schools of the city of Tunja, Boyacá during the weekly sessions of training.

2. Methodology

In the present study, took part 20 soccer coaches from the different sports training schools of the city of Tunja, Boyacá in the categories of children and youth attached to the league of this department. The study has a quantitative approach since a set of systematic stages in data collection was addressed to achieve the proposed objectives. "According to the scope, the research is descriptive, because the data collection was done in a single time allowing knowing and describing the main methods used by coaches in the Sports Training Schools of the city of Tunja Boyacá" [11].

The call for participation was made voluntarily informing previously about the study we would perform. The coaches were informed that the goals were strictly academic support for the educational and research work. The applied survey was designed as a result of the experience and of a rigorous bibliographic revision referring to the teachings that are taught in the Schools and sports training clubs in football. In the same way, the survey was modeled and validated using closed and open semi-structured questions techniques through the contributions of professors and football experts of Tunja, Boyacá.

The survey presented 3 moments for its structuring. The first was called General Data or characteristics of the School or club, where information was collected about name or business name of the school of sports training, year of creation, number of coaches, level of academic training, levels of sports training, schedule, number of children that make up the soccer school. As a second part of the survey, it was called technical data of soccer which is a soccer training session and how many parts it is composed of, which develops in each one, what a sports football warm-up is, what the physical skills and physical abilities are, what the technical-tactical sports of soccer are. The third part of the survey was called teaching methodology for soccer what methods do you use for teaching-learning? of soccer, knowing the definition of each method, what exercises it develops for the analytical, global and mixed methods, hoe frequent each method is used.

The previous aspects are examples that the survey presents for the respective data collection.

3. Results

The results that were obtained through the sports survey applied to coaches of the children and youth categories of the Sports Training Schools of the city of Tunja Boyacá, were treated and processed by means of statistical tools.

For the description of the results, the three categories established in the Survey Instrument were taken into account, taking as reference and importance for the present study the category called General Data or characteristics of the School or club, 1 item was taken into account, level of academic training of the Trainer, of the category Technical data, 4 items were taken into account, Knowledge of the trainer about a training session and its parts, Knowledge of technique and tactics, and finally the category called Methodology of teaching in soccer, 3 items were taken into account. Know and uses the teaching method in football Analytical, know and uses the global teaching method in football, know and use the teaching method in mixed soccer.

Table 1. Level of academic training of the football coach in areas at the end.

Students in sports	Professional in sports	Bachelor of Physical Education Recreation and Sports	Certified Trainer	No Level of Training
8	2	7	2	1

In Table 1 we can see the training level of 20 coaches evaluated. It is determined that there are coaches who exercise their work in soccer with different academic backgrounds from the level of not having a level of professional or empirical training in areas related of sports to the level of the licensed and certified trainer to guide the teaching processes learning soccer in children and youth categories.

In Table 1 we realize that, 20 soccer coaches were evaluated in terms of their level of academic training regarding soccer sport or related areas, we found that 8 coaches are studying in an academic training field, which corresponds to 40% of the population that guides the teaching-learning processes in the Sports Training Schools and Clubs of Tunja. In the same way, we find 2 coaches with a Professional academic level in sports which corresponds to 10%. Then we see that 7 coaches have a degree in Physical Education, Recreation and Sports which represents 35% of the population. There are also 2 certified trainers, which corresponds to 10%. Finally, we can see that there is 1 Coach that guides the teaching-learning processes without any academic training related to football or related to other sports areas which represents 5% of the 20 Coaches evaluated in Tunja.

It is concluded that almost the entire population has levels of certified academic training to work with this type of children and youth categories in the sport of soccer.

Table 2. Technical data of the sport.

Items	Yes	No
Learn about the training session	20	
Know the parts of a soccer training session	17	3
Know the sports technique and how it is composed	14	6
Know the sports tactics and how it is composed	15	5

In Table 2 we find the results obtained from the category Technical data of the sport, whose surveyed items are of the degree of knowledge that the coaches have about the teachings-lessons they teach in the Sports Training Schools and Sports Soccer Clubs. It can be seen that most of the coaches have knowledge bases and that the training aspects are carried out with reference to the items that make up this category of the applied instrument.

In Table 2 shows the percentages of the different items that make up category two of the instrument called technical data. We can see that the 20 coaches have knowledge about what a soccer training session is, which 100% is. Continuing we see that in the second item there are 3 coaches who do not know the parts of a soccer training session which is equivalent to 15% of the studied population. The remaining 17 are equivalent to 85% have great knowledge of the divisions and components of a football training session in each level or category of sports training of the participants. In item three, referring to the aspects of the technique and its components we observed that 14 coaches equivalent to 85% of the total surveyed demonstrate knowledge of the sports technique and the technical foundations of the soccer player. The remaining 6 coaches that are equivalent to 15% have a hard time having a clarity of the aspects that make up the technique; movements without and with the ball. Finally, in the fourth item, referring to sports tactics, we determined that out of 20 coaches surveyed 15 corresponding to 80% have a good knowledge of the aspects that make up defensive and offensive tactics. The remaining 5 coaches equivalent to 20% have low knowledge in the previous aspects.

Table 3. Teaching methods in soccer.

Items	Yes	No
Know and use the analytical method	14	6
Know and use the global method	16	4
Know and use the mixed method	14	6

In Table 3 describes the items that make up the category three of the instrument called teaching methods in soccer. As we observed, we found that of the 20 coaches surveyed, most of them know, define and use the analytical, global and mixed teaching methods used in the teaching of soccer. These coaches use these teaching-learning methods in a planned, orderly and sequenced manner of the specific exercises of each aspect of the technique and tactics. The coaches who answer negatively in the survey do not develop their training sessions in a very generic and empirical way without having a control, systematic and sequenced order in the performance of the teaching-learning exercises of their athletes.

In Table 3 we can see the three types of soccer teaching methods which the 20 Coaches of this sport belonging to Sports Training Schools in the city of Tunja Boyacá were surveyed. We find in the item one on the knowledge and use of the Analytical Method that 14 coaches corresponding to 70% of the established sample possess and master knowledge about how analytical teaching methods and the respective methodologies of exercises used to define and use are used to learn the different sports techniques and tactics of soccer. The remaining 6 correspond to 30% who affirm not to have a clear idea in the use and function that this analytical method fulfills. Next, in item two; regarding knowledge and use of the Global Method, we observe that this type of education is the one with the highest percentage of knowledge and application shown by the population. 16 Coaches corresponding to 80% argue that this type of training allows to be more effective for the player when learning why real plays are presented similar to the games that must be quickly solved by applying quickly the execution of the technique and the decision of the tactics to assimilate a good learning of the aspects that make up soccer. The remaining 4 trainers equivalent to 20% answer that they do not use this type of teaching methods in the training of athletes due to the little knowledge and information of the variability of exercises that can be done and that is worked in this method of teaching for soccer. Finally, we realize that the third item knowledge and use of the mixed method show that 14 coaches corresponding to 70% know how to apply the combination of analytical and global methods for teaching soccer in the children and youth categories. The remaining 6 are the equivalents to 30% are the ones who do not perform this type of mixed method because they find it difficult to integrate the combination of global analytical and global exercises. In addition, they lack knowledge and procedures that make up this type of methodologies for soccer training.

4. Conclusions

In the context of team sports, soccer occupies one of the first places that children, young people, and adults practice from different points of view and approaches.

In this way in the present research work, the main objective was to know and describe the methods of teaching football that is used by coaches to train athletes in children and youth categories in the Sports Training Schools in Tunja, Boyacá during the weekly training sessions.

It was possible to corroborate through the instrument applied, survey, that most of the evaluated soccer coaches of the sports training schools of Tunja, Boyacá, have a great knowledge of the methods that exist in the specialized literature to guide the training of children and youth during the training weeks.

It was found that the methods most applied by soccer coaches in Tunja are the global and analytical methods worked out in a separate way each for teaching-learning the technical and tactical aspects regarding soccer.

In the present study, it is revealed through the obtained results that the trainers of the city of Tunja, use and use more for their training the Global Method. However, the coaches in certain days of training emphasize the analytical method in children's categories for the work of the technique and emphasis to the mixed method for the work of the tactics in juvenile categories using real play of the game that is presented to the child or player inside the field in a training for the competition.

Finally, this study shows that the global methodology is the most frequented by soccer coaches in the children's and youth categories of the city of Tunja, but we can make the judgment that most coaches believe that training is more convenient than they are carried out by combining several methodologies for the teaching of sports disciplines.

References

[1] Pacheco R 2004 *La enseñanza y entrenamiento del futbol 7* (Barcelona: Paidotribo)
[2] Chirosa L J, Ponce F, and Chirosa I J 2003 El efecto de dos formas de aplicación de la técnica de enseñanza mediante indagación o búsqueda en fútbol base, sobre las capacidades condicionales y los medios técnico-tácticos individuales de la conducción y el control *II Congreso Mundial de Ciencias de la Actividad Física y el Deporte* (Granada: Universidad de Granada)
[3] Harvey S 2003 Teaching games for understanding: A study of U19 college soccer players improvement in game performance using the game performance assessment instrument *Segunda Conferencia Internacional: Enseñanza del deporte y educación física para la comprensión* (Australia: Universidad de Melbourne)
[4] Hayes S, Horn R, Hodges M, Scott M, and Williams M 2005 The relative effects of demonstrations and outcome information in the teaching of novel motor skills *Congreso Mundial de Ciencia y Futbol Lisboa* (Madrid. Gymnos)
[5] Sanchez J, Molinero O and Yague J 2012 Incidencia de dos metodologías de entrenamiento aprendizaje sobre la técnica individual de futbolistas de 6 a 10 años de edad *Retos: Nuevas Tendencias en Educación Física Deporte y Recreación* **22** 29-32
[6] Mitchell S, Griffin L and Oslin J 1995 An analysis of two instructional approaches to teaching invasion games *Research quarterly for exercise and sport* **66(A)** 64
[7] Ponce F 2006 *La eficacia de la utilización de una técnica de enseñanza mediante indagación o búsqueda en la mejora de diferentes aspectos funcionales, técnico-tácticos, decisionales y motivacionales, en futbolistas de categoría alevín, frente a una intervención tradicional y su transferencia a la competición* (Granada: Universidad de Granada)
[8] García J A 2001 *Adquisición de la competencia para el deporte en la infancia: el papel del conocimiento y la comprensión en la toma de decisiones en balonmano* (Badajoz: Universidad de Extremadura)
[9] Sánchez Bañuelos F 1986 El deporte como medio formativo en el ámbito escolar Ed. Blázquez D *La iniciación deportiva en el deporte escolar* (Barcelona: Inde)
[10] Sans Torrelles A, and Frattarola Alcaraz C 2006 *Manual para la organización y el entrenamiento en las escuelas de fútbol* (Barcelona: Paidotribo)
[11] Hernández S, Fernández C, and Baptista P 2004 *Metodología de la investigación* (México: Mcgrawhill)
[12] Merchán R 2010 *Nivel de eficacia técnica del jugador de futbol base* (Pamplona: Universidad de Pamplona)

IOP Conf. Series: Journal of Physics: Conf. Series **1161** (2019) 012009 doi:10.1088/1742-6596/1161/1/012009

The pedagogical competences and their importance in virtual physical education

J O Hernández-Gamboa[1], J L Vera-Rivera[2] and A J Ortega-Parra[1]
[1] Universidad de Pamplona, Pamplona, Colombia
[2] Universidad del Valle, Cali, Colombia

E-mail: dedufisica@unipamplona.edu.co

Abstract. Pedagogical competences have been imposed in relation to the development of face-to-face and virtual education policies in the world. In Colombia, the competency-based curriculum has been used at all levels of the Education System. However, there are shortcomings in the training of teachers in regard to the implementation and execution of different activities that allow the full development of the school and university environment. The research focuses on the pedagogical competences of the teacher of Physical Education at the Education level. It is a journey that took place, taking into account the orientations of qualitative research, under the hermeneutical paradigm and the ethnographic method. Likewise, the observation techniques were used. For the theorization process, the authors' approaches were taken into account, who proposes stages or levels for the development of a research: categorization, structuring, contrasting and theorizing. At the end of the research, ponders were made on the harmonic relationship in Physical Education, where the achievement of the objectives depends on the pedagogical skills assumed by teachers with teamwork, based on values, motor skills, sports, playful and other expressions of the movement seeking the fullness of the spirit in the body-soul relationship. Key words: Pedagogical competences, Face-to-face and virtual Physical Education.

1. Introduction

This article is the product of a literature review, documentary on the formation of pedagogical skills, under the humanistic approach to which the work done merges with teachers in Physical Education area in educational centers in the city of Pamplona, Norte de Santander, Colombia. It is expected that the teacher is the counselor of the process and the student becomes the main actor of educational action.

It should be noted that one of the most important stages within what corresponds to the holistic training that the human being should receive within the educational process at face-to-face and distance level, lies essentially in Education. At this level, the cognitive study of the physical and psychological abilities of the young or adult child prevails. It is a stage where creativity can increase, instill and train in values to achieve the full development of personality, under criteria of love, responsibility, honesty, work and creativity.

The pedagogical skills must constitute the fundamental motor in the integral development of human being. Physical education is a priority area to achieve this goal, since it is made by a compendium of activities that are established in a harmonic and holistic way in each action of being human, within its process of regular formation and daily life, with such naturalness that it has great influence on life and body.

IOP Conf. Series: Journal of Physics: Conf. Series **1161** (2019) 012009 doi:10.1088/1742-6596/1161/1/012009

The Physical Education in all the levels of the educative system (Preschool, basic, average and diversified and until the university stage), includes two vital aspects for its education: Body and Movement, which are constituted in two fundamental areas for the educational practice.

Physical Education aims as stated by [1] Fat (2005): "the formation of the human being", maintaining as a priority task, "the achievement of the body entity".

Seen in this way and with special emphasis on the ontology of the object of study, the definition of the [2] UNESCO (1999) it states what pedagogical skill is: "the set of socio-affective behaviors and cognitive abilities, psychological, sensory and motor, that allow to carry out adequately a performance, a function, a task "(s/p). The aim is to articulate skills from a humanistic perspective through the use of Physical Education in the stage of basic or secondary education where the teacher is able to promote social interaction in the educational process and where development takes place, when Internally, the cognitive process is controlled by integrating technology, virtualization and physical abilities into the mental schemes that contribute to meaningful learning. The contributions of [3] Vygotsky (2006) are of great importance in the educational process and, it is important to encourage Physical Education teachers, to investigate in order to improve the educational practice at distance, so that the teacher can be a dynamic agent committed to the true essence of education. In other words, achieve the transformation of society.

The learning proposed by [4] Ausubel, (2000) establishes that it is necessary to take into consideration the previous knowledge of the student, so that they serve as a bridge in the abstraction to which the programmatic contents are usually subjected.

2. Method

[5] Platón, (1988). and [6] Mulder (2007) points out that in the sixteenth century the concept was already recognized in English, French and Dutch; The use of the words competence and competence in Western Europe dates from the same period. So, it is important to make clear that the concept of competence has a broad history, so it is not surprising that being professionally competent means being sufficiently capable and able to perform certain tasks. It stands out that throughout the ages has been an aspiration, the search for skills to have control over an event or a specific technique.

According to [7] Argudín (2006), the term competence is derived from the Greek agon and agonistes, which indicates that one has been prepared to win in the Olympic competitions with the obligation to emerge victorious and therefore to appear in history. If an analysis of the definition is made from the etymological point of view, competence means being filled with capacities to face the challenges. In the same way it consists of avoiding obstacles and overcoming them. Action that impacts in depth with the purpose pursued through education.

In this sense, competence implies according to [8] Bustamante (2002): "The idea of an active and complex mind and therefore that of a productive subject". Hence, it can be said that competence is to forge a goal and be responsible for consciously generating the knowledge that it receives, processes and reproduces, based on experience and what comes from the context or environment to which it is circumscribed.

[9] Montenegro (2005), referring to the concept of competence, affirms that it is very broad, and defines it as follows: "being competent is knowing how to make and knowing how to act, understanding what is being done, understanding how to act, assuming in a responsible way the implications and consequences of the actions carried out and transforming the contexts in favor of human development".

[10] Tobón (2004), states that the origin of competences lies in Greek philosophy, due to the nature and method of work of the thinkers of that time. The subjects were approached by the Greek philosophers from contextualized problems that interrogated reality, being and man in an articulated way.

[11] Argudín (2006) states in this regard that: "Man faces radical changes that forces him to understand that now more than at any other time, he must meditate on his personhood and build his way". A path that includes a wealth of knowledge and skills that give him security to live in a system that qualifies skills and abilities. That is, it requires a life plan with clear goals.

2.1. Levels of competence in Education and especially Physical Education from distance
Competencies have different levels of development and these are made evident through performances, actions and behaviors. For students to develop them, it has to be through different levels of proficiency, and citing [12] Sánchez (2012) states that: "They are the sequence of learning with which the student reaches the completion of each competition".

In Colombia, as [13] Unell (1997) puts it: "There has been a set of pedagogical and didactic experiences, which have gone hand in hand with the evolution in the learning of motor skills and abilities by students in all the levels of the education system". In the same way, it is affirmed that the degree of maturity in the development and progress towards more complex systems or of better quality in terms of the realization of skills and abilities has also been taken into consideration.

It is important to delve from the concept of virtual Physical Education in its different meanings, which are always oriented to a historical moment given by the dominant conception of the body, and the knowledge that must be applied for the coordinated realization of movements, abilities and capabilities such as it affirms [14] Contreras (1998) "the idea of the human body has conditioned the concept of Physical Education in history". Therefore, Physical Education has acquired different meanings according to the concept of body, from an ontological perspective depending on the time of society.

In these terms, it should be noted that the specific competences of Physical Education, Recreation and Sports, which have been raised by the Ministry of Education, from an integral approach of the human being, are referred to:

2.2. Motor competence
This competition according to the [15] Ministry of Education of Colombia (2010): is understood as the construction of an autonomous body that gives meaning to the development of motor skills, physical abilities and movement techniques reflected in knowledge and skills useful to meet the crop personnel and the ever-changing demands of the environment with creativity and efficiency.

2.3. Expressive body competition
This competence as proposed by the Ministry of Education of Colombia (2010): addresses all the knowledge that the student has about himself (ideas, feelings, emotions), techniques to channel emotionality (release tensions, overcome fears, accept your body), of bodily and communicative availability with others, through the expression and possible, flexible, efficient and creative representation of gestures, postures, space, time and intensities. In the same vein, [16] Begoña (2005) points out that:

The development of the corporal expression in the school must take into account: to feel the body, the corporal expression and the dramatic language, from some theoretical foundations as they are the cognitive capacity, the physical or energetic capacity, the capacity of movement and the expressive capacity".

2.4. The axiological competence of the body
This competence is understood as a set of culturally determined values and are considered vital in the development and social coexistence. The knowledge and techniques acquired through physical and recreational activities, also contribute to build a lifestyle oriented to their care and preservation, as a necessary condition for the understanding of those social values and respect for customs, traditions and the environment. With regard to this competence, the tendency is to try to achieve the integral formation of the student from the corporal, considering it from a humanistic approach, which is proper to the social sciences and where the axiological value of each activity must be exploited to have solid basis that are generated from the context, where teamwork is fundamental to achieve the objectives, from the preservation of friendship, solidarity, responsibility, respect, tolerance, honesty which are virtues that will contribute to the formation of integral citizens that will give significant contributions in the construction of a better society. A society in values.

IOP Conf. Series: Journal of Physics: Conf. Series **1161** (2019) 012009 doi:10.1088/1742-6596/1161/1/012009

It is important to emphasize that, for the success of the teaching process and the learning process, the work of all those involved is necessary. Therefore, collaborative work is necessary for the fulfillment of the objectives that both teachers and students seek to obtain. In this circumstance, this collaborative work is based essentially on the precepts disseminated by [17] Vygotsky (1995), who writes: "Learning awakens a variety of development processes that are able to operate only when the child interacts with other people and in collaboration with their classmates".

3. Conclusions

The virtual education must implement didactic guidelines that help the teacher to clarify doubts, in order to achieve the formation of autonomous students, capable of making decisions, and transfer the learning of the class to everyday life.

The appropriate use of science, technology and virtualization through the use of strategies and didactic resources can facilitate the achievement of objectives, since their absence or lack of knowledge can become in distracting factors of the goals established in the area.

References

[1] Graso A 2005 *Construyendo Identidad Corporal. La corporeidad escuchada* (Argentina: Novedades Educativas)

[2] UNESCO 1999 *World Declaration on Higher Education in the 21st Century: Vision y Action* Consulted on: http://www.unesco.org/education/educprog/wche/declaration_eng.htm

[3] Vygotsky L 2006 *Psicología del Arte* (Barcelona:Paidós)

[4] Ausubel D 2000 *Adquisición y retención del conocimiento* (Barcelona: Paidós)

[5] Platón 1988 *Diálogos República* (Madrid: Editorial Gredos)

[6] Mulder M Weigel T and Collings K 2007 El concepto de competencia en el desarrollo de la educación y formación profesional en algunos Estados miembros de la UE: un análisis crítico Profesorado, Revista de currículum y formación del profesorado **12(3)** 1-25

[7] Argudín Y 2006 *Educación basada en competencias* (México: Trillas)

[8] Bustamante G 2002 *El concepto de competencia una mirada interdisciplinar* (Colombia: Sociedad de Pedagogía)

[9] Montenegro I 2005 *Aprendizaje y desarrollo de las competencias* (Colombia: Editorial Magisterio)

[10] Tobón S 2004 *Formación Basada en Competencias* (Colombia: Editorial Magisterio)

[11] Argudín Y 2006 *Educación basada en competencias* (México: Trillas)

[12] Sánchez T 2012 *Renovación Curricular* (Chile: Universidad Católica de Temuco)

[13] Unell B 1997 *20 valores que usted puede transmitirles a sus hijos* (Bogotá: Editorial Norma).

[14] Contreras O 1998 D*idáctica de la Educación Física: Un enfoque Constructivista* (Barcelona: Inde)

[15] Ministerio de Educación de Colombia 2010 *Documento no. 15 Orientaciones Pedagógicas para la Educación Física, Recreación y Deporte* (Bogotá: Ministerio de Educación)

[16] Begoña R 2005 *Didáctica de la Expresión Corporal* (España: Inde)

[17] Vygotsky L 2006 *Psicología del Arte* (Barcelona:Pai dós)

IOP Conf. Series: Journal of Physics: Conf. Series **1161** (2019) 012010 doi:10.1088/1742-6596/1161/1/012010

Development of a virtual learning environment for the subject numerical methods under Moodle

A Becerra-Romero[1]**, M Díaz-Rodríguez**[2]**, and O A González-Estrada**[3]
[1]Departamento de Tecnología para el Diseño Industrial, Universidad de los Andes, Mérida, Venezuela
[2]Laboratorio MECABOT, Universidad de los Andes, Mérida, Venezuela
[3] GIEMA, Universidad Industrial de Santander, Bucaramanga, Colombia

Email: dmiguel@ula.ve

Abstract. Numerical Methods is a core subject for most graduate and undergraduate engineering curricula. The course topics are transversal to most of the applied science courses. That is, any specific topic can include a problem requiring a numerical solution. Thus, developing an improved educational strategy for this subject is of paramount importance. Virtual learning environments (VLE) offer a tool for improving students' learning outcomes. In this paper, we describe how we build a VLE for the subject numerical methods, based on Moodle. We use GeoGebra software and R programming language to develop virtual-learning-objects (VLO). To engage students, we propose a problem-solving activity based on movie action sequences. Each topic starts out with a movie clip that using physics principles, one can get to an equation requiring a numerical solution. For example, we explain the methods for solving ordinary differential equation by modelling a parachute sequence from a James Bond movie. The activity allows students to discuss whether the movie sequence is possible or not. Finally, we present preliminary feedback obtained on a graduate course. Also, we discuss the answers from an instructional design questionnaire applied to the course.

1. Introduction

Nowadays, it is almost taken for granted that any subject within curricula is taught taking advantages of virtual learning environments (VLEs). Even if the course follows traditional master class, the VLE helps as additional resource or at least it helps for managing the course content and marks. Moreover, VLEs are useful when teaching Science and Technology, in general because they permit the implementation of many learning objects such as: YouTube videos, podcast, presentation slides, programming codes, and links to other websites or animations showing in a dynamic way simulations and concepts very difficult to understand by the students [1]. A VLE or also known as learning management system (LMS) is the environment where students and professors participate, and the e-learning takes place [2]. The implementation of LMS has been proved to be a successful for Knowledge Management within organizations [3].

On the other hand, in most of the engineering degree curricula, Numerical Methods is traditionally taught following comprehensive textbooks such as [4]. The subject presents fundamental knowledge for most engineering programs at undergraduate and graduate level; thus, any strategy to enhance its understanding is valuable. In this sense, [5-6] reported that teaching numerical methods using VLE compared to traditional approaches presents a significant improvement in learning outcomes. [7]

developed a meeting point in Moodle, for students of Numerical Methods for Computer Science Engineering at the University of Malaga. Results showed that VLE enhances students' outcomes.

At Universidad de los Andes, Venezuela (ULA), the Coordinación General de Estudios Interactivos a Distancia (CEIDIS) (http://moodle.ula.ve/) have developed a Moodle platform for online courses such that it can serve as an alternative or complementary way of teaching, covering levels of undergraduate and graduate. The motivation of this paper is to showcase the VLE developed for teaching Numerical Methods. The VLE was put forward following instructional design concepts described by CEIDIS as well as the authors' own contributions. One of the novel aspects of the course is to capture student's attention by introducing a problem-solving activity based on movie action sequences. Each topic presents a movie clip that can be stated as an engineering problem requiring a numerical solution, *e.g.*, the motorbike jump sequence from Point Break movie. The jump can be described using a nonlinear equation admitting a numerical solution. So the equation from the movie clip is solved through the lesson using numerical methods. Also, for enhancing the teaching experience, several virtual learning objects (VLO) are selected from online resources or developed by the authors.

The paper is organized as follows, Section 2 introduces Numerical Methods course content. Section 3 describes the basic features of the Numerical Methods VLE. Section 4 shows the VLOs selected or developed for each course topic. Section 5 discusses the preliminary assessment of the course. Finally, conclusions and further works are presented.

2. Subject description

The subject Numerical Methods for Engineers provides a basic introduction to the methods available for the numerical solution of engineering problems. In order to implement the numerical solution, students develop programming skills. Software such as: Python, R, Matlab, C++, FORTRAN, can be used in laboratory sessions to learn how to coding while implemented a specific numerical method. The students take the course after the block of basic science knowledge: mathematics, calculus, physics, and before the block of subjects in which specific applied engineering knowledge are taught. Then, in subsequent courses students are able to implement numerical methods. For instance, in the Mechanical Engineering curricula at the Universidad de los Andes, Venezuela, the students can take the course in the fifth term (to obtain the degree the students have to course ten terms where each term last a semester). The topics included in the course are: solution of nonlinear equations and linear equations, data fitting through linear regression, interpolation, numerical integration, numerical differentiation, and solution of ordinary differential equations. The course is not intended to cover all numerical methods but rather to provide a basic skill set useful throughout their education.

As an example, one of the courses following numerical methods, we have the mechanism and machine theory course, specifically the position problem of a mechanism. When developing the equations related to the position problem, students often arrive at a set of nonlinear equations [8]. In a robotic course, the solution of the kinematic problem is related to solving a set of linear equations [9]. Further examples can be presented to highlight the importance of numerical methods which to our opinion is a transversal axis within engineering curricula.

At the end of the course, students are expected to understand and apply the methods for,

- Assessing the error or approximation of a numerical solution.
- Finding the numerical solution of a nonlinear equation.
- Finding the numerical solution of a linear equation.
- Finding the parameters of a linear model through linear regression.
- Computing the numerical integration of an equation.
- Computing the numerical derivative of an equation.
- Solving ordinary differential equations.

The course content is presented through six thematic blocks. The first block addresses the introduction putting emphasis on the importance of numerical methods, describing many examples where numerical methods can be applied. The second block describes methods for the solution of nonlinear equations, while the third block describes the solution of linear equations. The fourth block deals with the data fitting through linear regression, and how to model experimental data using linear models. The fifth block describes how to compute the numerical solution for differentiation or integration of equations. Finally, the last block refers to the solution of ordinary differential equations.

3. The virtual learning environment

We developed the structure of the VLE following the set of courses provided by CEIDIS, and the best practices on how to organize a Moodle course [10-12]. First, after studying the course's content, we decided to take the Moodle Topic format. We split the course content in 10 topics. Each topic covering a specific numerical method, and they will be visible to the student after finishing the corresponding assignment of the previous topic. The topic starts out by setting the goals, answering what are the competencies students will expect to learn. Then, it follows the presentation of the learning material. After that, the topic includes an activity section where the learning objects are presented. The topic ends by assessing the learned skill through an evaluation activity. The structure of the topic is similar to the CEIDIS basic design course structure,

- Basic Competencies
- Content
- Activity
- Evaluation

Figure 1 shows an example of each topic structure, the figure specifically presents the content of topic two which is related to the errors and approximations underlying in a numerical solution. The topic follows the Introduction and precedes the topic on solving nonlinear equations.

The topics in the VLE are organized as follows:

- Introduction (Block 1)
- Errors and approximation (Block 1)
- Bisection and fixed-point iteration methods for solving of nonlinear equations (Block 2)
- Newton-Raphson and Secant method for solving of nonlinear equations (Block 2)
- Simple Gauss elimination method for solving linear equations (Block 3)
- Gauss-Jordan and Gauss-Seidel method for solving linear equations (Block 3)
- Data fitting (Block 4)
- Numerical Interpolation (Block 5)
- Numerical differentiation (Block 5)
- Numerical solution of Ordinary Differential Equations (Block 6)

First, each topic starts by describing the competencies. Then, the learning material is introduced after showing a movie clip related to a problem requiring a numerical solution. For instance, the teacher can start topic 3 by playing the video of the high-altitude low-opening (HALO) parachute jump sequence from the James Bond movies Tomorrow Never Dies. The HALO jump can be modelled through a differential equation that in the evaluation section will be solved by a numerical method. Due to the paper length limitation, we refer readers to see (http://mecabot-ula.org/blog/ecuaciones-diferenciales-aplicadas-al-salto-halo-debond-tomorrow-never-dies/) to understand how mathematical equation for the movie sequence is developed. The topic include a forum section where students discuss whether the movie sequence is feasible or not, which engages his/her attention in understanding the problem and its solution. After presenting the motivation activity, each topic includes readings taken from the book Numerical Methods for Engineers [4], Figure 1 presents the structure of topic one, the structure is shown in the language in which the course was developed.

IOP Conf. Series: Journal of Physics: Conf. Series **1161** (2019) 012010 doi:10.1088/1742-6596/1161/1/012010

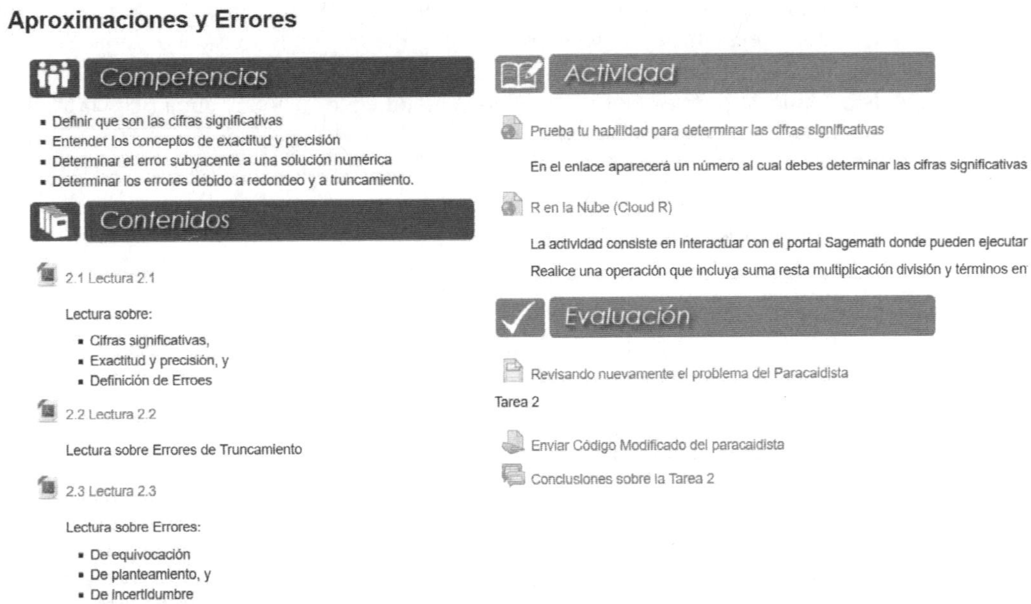

Figure 1. Screen-shot of topic two: Errors and Approximations (text in Spanish).

The first topic includes three specific readings. Then, the topic presents activities so that the student can put in practices the knowledge presented in the content section. In the activity section presents the learning objects designed to consolidate the learning outcomes. Then, the evaluation section includes a forum where student can discuss the results found in the evaluation activity. The discussion section is open to any student so they can interact in a chat format.

Due to the fact that each block presents not only similarities but also differences, we develop for each block different learning objects which are presented in the following section.

4. Virtual learning objects

The similarity in all blocks is given by the fact the student must implement the corresponding method learned in the topic. Thus, a learning objective transversal to all the topic should be an activity based on developing programming skills. In our case, we take advantages of the R programming language; however, any high-level programming language can be selected for such activities.

4.1. Learning objects based on R programming language

We developed the programming objects in R based on the fact that is a free software available under the GNU General Public License, also it has high statistical and graphics tools that we want to implement in the Mechanical Engineering curriculum. When working at home, the students can combine R package with the graphical front-end RStudio which is one of the most notably available free and open-source Integrated Development Environment (IDE). How to download and install R and RStudio is presented in topic one. Before sending the evaluation assignments, we demand the student to first test the code on SageMathCell (http://sagecell.sagemath.org/) which is a web interface software system where you can run code in python, sage, R, and many other programming languages. In this way, we are able to evaluate the students by asking them to modify or implement a set of code in R-language. Then, they can copy and paste the code on the Moodle activity task tool. Thus, we avoid the issue of uploading or sending any file.

To exemplify the use of R as a learning object, the code for the topic 3 is presented in Appendix I. The example presents the code for solving a nonlinear equations through the Bisection method. The nonlinear equation is taken from the engineering problem presented in Ref. [13]. In the activity, students can copy and paste the code in his/her one IDE. Then, they can run it to interact with the results by

IOP Conf. Series: Journal of Physics: Conf. Series **1161** (2019) 012010 doi:10.1088/1742-6596/1161/1/012010

modifying parameters of the numerical method, for instance changing the tolerance threshold parameters (`tol`). Finally, students can reuse the code when performing the evaluation task. For topic three, the task is based on solving the equation for projectile motion applied to the bike jump sequence presented in the movie Point Break (see, http://mecabot-ula.org/blog/aprendiendo-metodos-numericos-por-medio-de-peliculas/).

4.2. Learning object based on GeoGebra

We use GeoGebra software for developing some of the learning objects. Objects created using GeoGebra have the potential that students can interact with the parameters of the numerical method being taught in a free and dynamic way, which helps to visualize the different behaviors through graphs. Students can estimate by their own guesses, ideas and conclusions, achieving a more lasting and significant learning through manipulation, exploration and experimentation [14]. The core of the method for blocks two and five of the course can be better explained using a graphical solution. Those blocks are suitable to develop a solution in GeoGebra making visual the interpretation of the method.

As an example, Figure 2 shows the solution of a nonlinear equation through the fixed-point iteration method.

The use of GeoGebra presents the advantage that we can reuse lots of resources and objects already on the web. Thus, we did not have to develop new objects, remember that reuse of object is one of the advantages when developing web-based courses. Instead, we search in the Classroom Resources page of GeoGebra (https://www.geogebra.org/materials), which also provide the capability of sharing the object such that we can embed the object within the specific topic in the Moodle VLE.

The list of objects, topic and link to interact with the object is listed below,

- Topic 3, Bisection method (https://www.geogebra.org/material/show/id/eR2qraTz)
- Topic 4, Fixed-point iteration method (https://www.geogebra.org/material/show/id/BFnKV77d)
- Topic 4, Newton-Raphson method (https://www.geogebra.org/material/show/id/GHuszVNm)
- Topic 8, Numerical integration (https://www.geogebra.org/m/R9GB9tfu)

4.3. Additional learning objects

The web offers lots of material we can reuse for different purposes, *e.g.*, block one deals with the fact of how the significant figures are relevant to the numerical solution and the error. After conducting a web search, we found out an application to play and learn about significant figures (see http://www.educaplus.org/formularios/cifrassignicativas.html). The app randomly presents a number and the user has to answer how many significant figures have the number displayed. In this way, the students self-evaluate his/her ability to identified significant figures. The activity section of topic one includes a link to the app, see Fig. 1, the app can serve as a learning object.

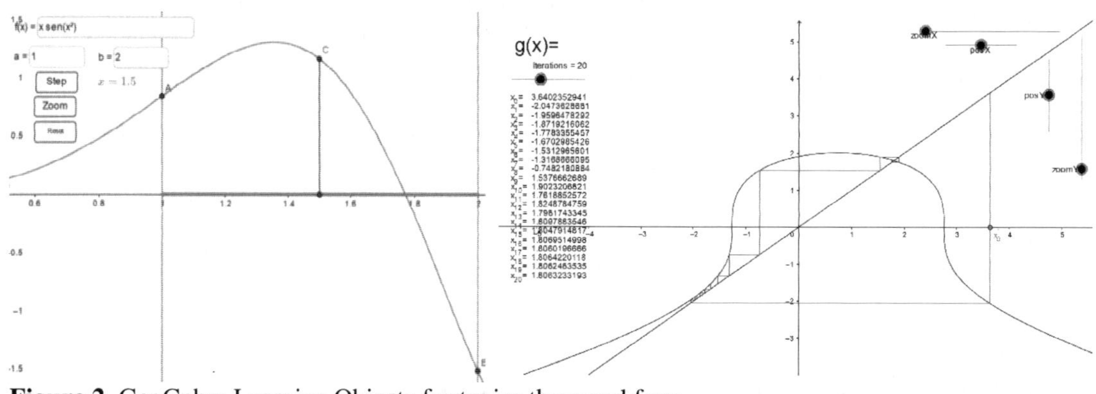

Figure 2. GeoGebra Learning Objects for topics three and four.

5. Preliminary assessment of the course

The course was taught to graduate students following a hybrid approach. That is, both online and face-to-face components were implemented. The course normally is based on 4 hours face-to-face week lectures. In the hybrid approach, students interact in classroom only 2 hours. Seven students, which scored above 80 scale from 0 to 100 points, participated. Each assignment was performed in due time. The activities showed positive feedback, and lots of discussions were presented in the forum section. Participants gave a mean score of 4.4, scale from 1 to 5, when asked for the activity regarding the movie sequence. In some movie clips, the students agreed clips were feasible, based on their own numbers, and while others found out the clip was unfeasible. However, one of the students claimed that at the end it is a movie, so action sequences even if they are not feasible as long as they are good enough to entertain they are worth.

In addition, the VLE was evaluated by an expert under the subject on analysis of an instructional design of the Master Degree in Educación Informática y Diseño Instructional, at the Universidad de los Andes. Overall, the expert provided positive feedback. With respect to the cognitive aspects, the VLE was developed following a constructivist approach. The comprehension of the learning material is based on cognitive operations such as logic operation and reasoning. The VLE promotes active participation through forums using written language, the resolution of the assigned activities, review of the material in a visual way or through podcasts, reading skills, the searching of information for choice concepts, similar applications, observation and discussion of results, and also the VLE presents results graphically in on-line simulator.

With respect to the behavioural aspects, the VLE presents a sequential order to each assigned activity and questions so that the students achieve the competencies that will enable them to provide solutions to the problems presented. Students have to repeat the step shown through tutorials, *e.g.*, to install the R program used during the programming assignments. The VLE promotes the interaction with their peers through the chat platform. To enhance the student mental processes, the VLE presents the instructions through readings in which the teacher developed the particular activity. Thus, students can take notes, view and listen to the tutorial videos on their computer and install the program to use, participate in written in forums either by giving their perception or asking questions, clicking on the documents to download. Students have to complete written activities, practical exercises in the R-program and write the conclusions based on the results.

With respect to the constructive aspect, student are expected to ask questions on the forum section, comment on what they have learned relying on previous knowledge, provide solutions to practical exercises, build numerical calculation using RStudio, develop numerical solutions to engineering problems, interpret the results obtained from a numerical solution, and recognize that numerical methods represent approximate solutions.

Finally, the expert recommended that the VLE has to extend the constructive approach from individual learning as it is right now for an approach based on collaborative learning. For instance, the VLE can include a project where each student can write a part of code that combined together they can solve the task at hand. In addition, the actual design of the VLE only allow the construction of knowledge within the VLE, they recommend building knowledge outside the VLE.

6. Conclusions

In this paper, we have showcased a VLE developed in Moodle for teaching Numerical Methods. The course structure, and the learning objects designed for the course have been also presented. In addition, we have discussed the preliminary assessment of the course based on its implementation in a graduate course, and also from a questionnaire applied for an expert on instructional design. The developed VLE represents a tool for e-learning, blended learning or flipping classroom strategies. Currently, the course is being evaluated on undergraduate courses, future works will provide the assessment from students' feedback.

IOP Conf. Series: Journal of Physics: Conf. Series **1161** (2019) 012010 doi:10.1088/1742-6596/1161/1/012010

References

[1] Martín-Blas T and Serrano-Fernández A 2009 *Comp. Educ.* **52** 35

[2] Weller M 2007 *Virtual Learning Environments* (London and New York: Routledge)

[3] Lache L, León A, Bravo E, Becerra L and Forero D *Rev. UIS Ing.* **15** 27

[4] Chapra S C and Canale R P 2006 *Numerical Methods For Engineers* (New York: McGraw-Hill Higher Education)

[5] Handayanto A, Herman T and Fatimah S 2017 *J. Phys.: Conf. Ser.* **895** 012068

[6] Handayanto A, Supandi S and Ariyanto L 2018 *J. Phys.: Conf. Ser.* **1013** 012128

[7] Mora-Bonilla A and Mérida-Casermeiro E 2011 *Rev. Med. y Educ.* **38** 201

[8] Mercado D M, Murgas G H, MCkinley J R and González J D 2015 *Rev. UIS Ing.* **14** 59

[9] Quintero-Riaza HF, Mejía-Calderon LA, Holguin G, Díaz-Rodríguez M, Herrera-López MJ, Mesa CA, Daraviña-Peña G 2018 *Manipuladores paralelos: Síntesis, análisis y aplicaciones* (Pereira: Universidad Tecnolíogica de Pereira)

[10] Nash S S and Moore M 2014 *Moodle course design best practices* (Birmingham-Mumbai: Packt Publishing Ltd)

[11] Gallardo-Pérez H de J 2014 El modelo pedagógico dialógico crítico en la educación *Respuestas* **19** 81–92

[12] Peñarando Guevara L A, Reyes Vera J M and Machuca Villegas L E 2013 Modelo de un sistema multiagente para proveer ayudas adaptativas en ambientes virtuales de aprendizaje *Respuestas* **18** 102–11

[13] Ramirez-Matheus A, Díaz-Rodríguez M and González-Estrada O A 2017 *Rev. UIS Ing.* **16** 197

[14] Allan C, Parra S and Martins A 2017 *TE & ET* **20** 51

Appendix I. Example of a R-code learning object.

```
#parameters
a=0.2;#[m]
b=1.2;#[m]
c=0.75;#[mm]
d=1;#[m]
K1=d/a;
K2=d/c;
K3=(a^2-b^2+c^2+d^2)/(2*a*c);
theta_2=45*pi/180
f=function(x)K1*cos(x)-K2*cos(theta_2)
+K3-cos(theta_2-x);
curve(f,0,2*pi,n=2001)
abline(h=0,v=0,col="gray60")
a=1;b=2
tol=1e-5
iter=0
while(abs(b-a)*0.5>tol){
c=(a+b)/2
if(f(a)*f(c)<0){
b=c
}else{
a=c
}
iter=iter+1
}
print('Resultados')
print('*****************************')
print('theta_4 Iter_Error')
print('*****************************')
c((a+b)/2*180/pi,iter,(b-a))
```

IOP Conf. Series: Journal of Physics: Conf. Series **1161** (2019) 012011 doi:10.1088/1742-6596/1161/1/012011

Strategies for the teaching-learning experiences in the engineering dynamics course based on the information and communication technologies

D F Villegas[1], H G Sanchez[1] and C A Riberos[1]
[1] Grupo de Investigación en Energía y Medio Ambiente (GIEMA), Universidad Industrial de Santander, Bucaramanga Colombia

Email: dfvilleg@uis.edu.co, hgsanche@uis.edu.co

Abstract. Technological developments in almost all sectors of industry and/or society has permitted to implement them even in universities by means of it is called ICT. This commits the universities to modify their educational approaches towards those that allow professionals be trained according to what the globalized industry demands today. In addition, it has been taking advantage of the virtual platforms in support of the teaching/learning experiences for the undergraduate courses that facilitate some activities like either the implementation of software of engineering or the student participation in forums or the revision of some concepts which can be impossible to understand in the traditional way of teaching. The Universidad Industrial de Santander (UIS) in Bucaramanga, Colombia, is looking for building a future with a continuous improvement, has undertaken several institutional projects in search of being at the forefront of the pedagogical and social requirements; This study sought to impact the current methods of teaching, and thus achieve a work in synergy with all the university that allows to show the UIS as an entity that continues to promote integrity and quality professionals. This project was divided into two stages. The first included the selection and thematic organization of the course, creating the virtual classroom in the Moodle platform and developing activities and tools that will be used there. The second phase was implementation. Always prioritizing skills training that improves quality and allows the future professional to adapt and meet the requirements demanded by the industry today.

1. Introduction

Currently the world is developing at a rapid pace, generating continuously great social, cultural, economic and technological changes. Technology and education are closely related each other nowadays, and the students are expose to a technologized culture [1]. Therefore, it makes evident the need to train professionals who should be not only adapt, but also anticipating to these changes. In other words, professionals can propose right solutions to these challenges [2]. Higher education has the responsibility to enhance this process, so it involves a great challenge for universities and especially in developing countries. Several studies have been done in Colombia in order to incorporates ICT experience in the classroom [3,4].

The design and implementation of new models, strategies and methodologies to improve the teaching-learning process has been the focus of attention of many researchers in the area. These studies have reported results obtained from the application of these new processes and tools. Some interesting work has done by Cinar and Torenli [5] where it was used online courses based on student expectations;

Redondo et al [6] used virtual and augmented reality to improve teaching-learning processes in an architecture program; Hauge y Riedel [7] used simulation of games in engineering and manufacturing teaching processes; and Farías et al [8] have implemented remote virtual laboratories to support engineering learning processes. It was seen a tendency in the previous studies to use ICT as a tool to innovate in the teaching-learning processes.

The use of information and communication technologies (ICT) have shown that it can improve not only the teaching-learning process, but also the performance in the working life of the professional [9]. In addition, it has been identified that it improves the teaching abilities of the instructor, changing the educational structure and contributing to the improvement of the quality education [10].

In this scene, the Universidad Industrial de Santander is looking for building a future with a continuous improvement and it has undertaken several institutional projects in search of being at the forefront of the pedagogical and social requirements. This study sought to impact the current methods of teaching, and thus achieve a work in synergy with all the university that allows to show the UIS as an entity that continues to promote integrity and quality professionals.

This project was divided into two stages: 1) it includes the selection and organization of the subject by themes in such a way that it allows to design a virtual classroom in the open-source course management system named Moodle (Moodle™) and development of the first activities and tools that are used there. 2) it is the implementation which focuses in the skill training that improves quality and allows the future professional to adapt and meet the requirements demanded by the industry today.

Engineering Dynamics (ED) is one of the fundamental courses in a Mechanical Engineering program. In addition, ED at UIS is one of the most failed courses in the program. It is thought that there are several causes on the difficulty of the students to get the complex concepts in an easy way. The ICT use have become in a need as a teaching-learning strategy because of the increasing technology during the last two decades. It has been demonstrated that ICT use improve the teaching-learning because of the students feel more motivated, have more interest on learning in such a way that they are involved in an active learning. Nowadays, students have higher IQ but the read less. Therefore, the written information should be short and contain more audiovisual resources in order to catch the attention of students more easily [11].

2. Methodology

Three groups of the Engineering Dynamics course were used for this study. Each course had a different teacher (the three authors of this article) which allow not to cut to the bias of the sample due to different teaching ways. Each course had around thirty students in average. The course was designed using Moodle (Moodle™) which the course management system used at Universidad Industrial de Santander. The design has into account the student outcomes which can be seen in Table 1 and the strategies proposed for the three faculty members teaching the ED course.

2.1. Course design using the course management system Moodle

Table 1 describes the strategies and activities based on: 1) the experiences of the authors in several years on teaching the course, 2) Moodle resources, and 3) several previous studies which implement education strategies [9,12]. On the other hand, the cognitive, attitudinal, and axiological competencies for the ED course are listed in Table 2.

2.2. Learning evaluation

The process for evaluating both the learning and the process of learning was carried out using different instruments like quizzes on-site and online, exams, assignments, essays among others. Table 3 lists the evaluations which were used to observe the student performance. For the implementation of online quizzes, a hundred and thirty questions were developed and the divided into the subjects of the course.

Table 1. The strategies and activities proposed and their relationship with the educational materials and Moodle resources.

Strategy	Activity	Educational Materials	Moodle Resources
Problem solution	Code development solve ED problems.	Programming Software: MATLAB	Assignment
Text composition	Essay about a subject in ED	Books, scientific articles, web sites	Bulletin board
Managing mistakes	Solutions of exams: identifying mistakes and correct them.	Copy of the exam solution	Assignment
Project Learning	Development of a specific subject in Engineering Dynamics	Engineering Dynamics textbooks, videos, and scientific articles	Evidence, portfolio
Online evaluation	Quizzes about Kinematic and Kinect of particles and rigid body	Presentations	Questionnaire
Research	Information in internet and library resources about a subject in the course.	Library resources, Websites	Wiki
Teaching	Teaching-learning process	Presentations, blackboard, and software for equation solving	URL, and labels
Student assignments	Solving assignments and upload the solution to the virtual room	Library resources, websites, computational engineering solvers	Assignment

Table 2. Cognitive, attitudinal, and axiological competencies for ED course withdraw from the syllabus based on learning competencies.

Student outcome	Description
1	Determine the position, velocity, acceleration and total travelled distance by a particle in order to determine the physical aspects of its movement.
2	Represent the movement of the particle in plots and interpret them.
3	Define the relations existing between forces, displacement, velocities, accelerations and mass of particles.
4	Explain the position, velocity, and acceleration of a rigid body.
5	Relate the forces acting over the rigid body and its effect over the shape and mass, in addition its movement in two dimensions.
6	Fulfill all commitments acquired at the beginning of the course with responsibility and ethics.

Table 3. List of type of evaluation for learning in ED course.

Evaluation type	Learning / learning evaluation	Resource
Quizzes about concept application and problem-solving procedures	Learning	Questionnaire
Quizzes about theoretical aspects	Learning	Questionnaire
Partial Exam	Learning	Written evaluation
Final Project	Learning	Evidence
Assignments and essays	Learning	Homework
Discussion about a subject in Engineering Dynamics	Learning	Open forum
Perception about using ICT as a support for learning	Learning process	Survey

3. Results

Moodle resources were known after several workshops offered at Universidad Industrial de Santander. The course was divided in six tabs according to the subjects: 1) A presentation of the course, 2) Kinematics of particles, 3) Kinetics of particles, 4) Kinematics of Rigid bodies, 5) Kinetics of rigid bodies, and 6) Kinetics of rigid bodies: Energy and Impact.

3.1. Implementation of the engineering dynamics course in the Moodle platform

All the strategies and activities proposed in Table 1 were implemented at the Moodle platform. There were several strategies that appears in each tab in Moodle as it can see in Figure 1. These are: A) a picture referring to each subject to gain the attention of the students, B) Wiki, C) Presentations for each class, D) Suggested computational problems, E) A case of study, F) Forum about concepts, H) Online quizzes, Part G) is optional.

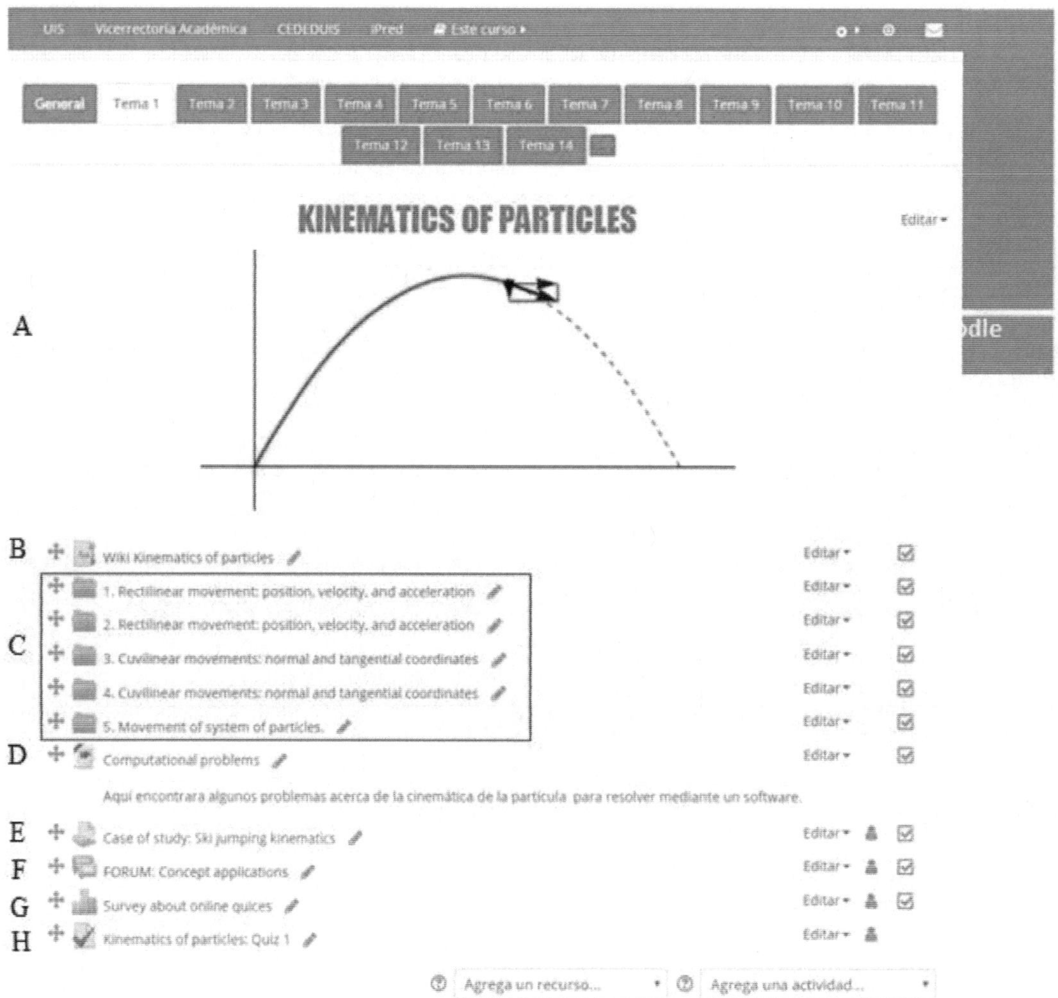

Figure 1. Print-screen of the subject “Kinematics of Particles” implemented in the Moodle Platform. I. It is shown several activities proposed in Table 1.

3.2. Online evaluation perception

Online quizzes represent an important tool because they allow a formal evaluation of the contents in a more practical way. A survey was applied to the student ($n = 20$) to have a perception of the students regarding this type of evaluation and compare it with the traditional way. The responses are shown in Table 4.

Table 4. Results of the survey for online quizzes.

Question	Responses			
Did you like online quizzes??			YES	NO
			85%	15%
Why did you like online quizzes??	Because we know the grade immediately	Because the test only evaluates answers	Because they are multiple selections	Because they are easy
	60%	5%	25%	10%
Would you like to take more online quizzes?			YES	NO
			65%	35%
Would like to take an online partial exam?		YES	NO	DEPEND ON
		30%	15%	55%
What the last question depends on?	Enough time to solve it	Difficult rate	Type of answer	Weight in the total grade
	70%	15%	5%	10%

Some interesting results are withdrawn from the survey. Most of the students liked to take online quizzes because the grade is given to them once they finish the test and they do not wait until teacher grade them manually. Some of them (25%) liked the online quizzes because the answer can be chosen from multiple answers. They also respond positively on the possibility on taking an online partial exam. However, time is the major concern (70%) on taking an online partial exam.

4. Conclusions

It was possible to develop the instructional design of the Engineering Dynamics course taking into consideration the training based on competencies with an inductive learning style. The learning object was developed for the 5 general themes of the ED course in Moodle by representing the fundamental pedagogical tool that justifies the improvement in the rate of approval of students, the disposition and motivation to take the course.

The object of learning developed allowed to see students participating in the construction of their own knowledge through the various tools they find in an autonomous way with the motivation to achieve all the competencies. In addition, the learning object greatly benefits teachers by generating a sequential, complete, unified and modern teaching methodology. Besides the benefit of an easy follow up to the teaching process because of a faster grading of some evaluation activities that Moodle synthesizes with tables and general diagrams of results.

On the other hand, the use of ICT in the teaching/learning process allows versatility in the way of presenting the contents and in the form of interaction of the student with them due to the different possibilities of presenting an idea or concept (videos, animations, diagrams, use of the web). It promotes collaborative learning between teacher-students and provides the teacher to continuously implement new technological tools which improve their process as an educator. In the same way, the teacher is also exposed to a teaching environment in which

Moodle has demonstrated to be one of the most important course management system that is available to higher education for the implementation of learning objects or simply to execute activities that complement the classes in the classroom, for example online questionnaires. It presents a simple and well-known interface for most students, facilitating a positive impact on the new methodology.

IOP Conf. Series: Journal of Physics: Conf. Series **1161** (2019) 012011 doi:10.1088/1742-6596/1161/1/012011

Future studies can be done in other courses of Mechanical Engineering program and also uses more elaborated online quizzes that allow to teachers to grade not only the answer but also the solution process in engineering problems.

Acknowledgments
The authors deeply thank to the Mechanical Engineering students at UIS whose took part in the survey, especially to Oscar Florez and Walter Valdivieso for their valuable help on this study.

References
[1] Rodriguez-Ruiz M 2011 Las TIC en la educación superior en Colombia *Rev. Unilatina* **6** 7–23
[2] Parrish D R 2016 Principles and a Model for advancing future-oriented and student-focused teaching and learning *Procedia - Soc. Behav. Sci.* **228** 311–5
[3] Moreno-Chaustre J J, Andrade-Sosa H H, García-González J R, Hernández-Pino U, Maestre-Góngora G P and López-Mólina G 2014 Modelo de evaluación para valorar el cambio en las prácticas docentes con TIC *Revista UIS Ingenierías* **13** 7–22
[4] Emiro L and Bravo C 2011 Uso de las TIC´s y especialmente del blended learning en la enseñanza universitaria *Rev. Educ. Y Desarro. Soc.* 151–60
[5] Cinar M and Torenli N 2010 Redesign online courses with student expectations: a case study with a new infrastructure *Procedia - Soc. Behav. Sci.* **9** 2013–6
[6] Redondo E, Fonseca D, Sánchez A and Navarro I 2013 New strategies using handheld augmented reality and mobile learning-teaching methodologies, in architecture and building engineering degrees *Procedia Comput. Sci.* **25** 52–61
[7] Hauge J B and Riedel J C K H 2012 evaluation of simulation games for teaching engineering and manufacturing *Procedia Comput. Sci.* **15** 210–20
[8] Farias G, Gomez-Estern F, De la Torre L, de la Peña D M, Sánchez C and Dormido S 2012 Enhancing virtual and remote labs to perform automatic evaluation *IFAC Proc. Vol.* **45** 276–81
[9] Nikolić V, Petković D, Denić N, Milovančević M and Gavrilović S 2019 Appraisal and review of e-learning and ICT systems in teaching process *Phys. A Stat. Mech. its Appl.* **513** 456–64
[10] Claro M, Salinas A, Cabello-Hutt T, San Martín E, Preiss D D, Valenzuela S and Jara I 2018 Teaching in a digital environment (TIDE): Defining and measuring teachers' capacity to develop students' digital information and communication skills *Comput. Educ.* **121** 162–74
[11] Twenge J M 2009 Generational changes and their impact in the classroom: Teaching Generation Me *Med. Educ.* **43(5)** 398-405
[12] Boix Tomas R 1995 *Estrategias y recursos didácticos en la Escuela Rural* (Barcelona, Universidad de Barcelona y GRAO de Serveis Pedagogics)

IOP Conf. Series: Journal of Physics: Conf. Series **1161** (2019) 012012 doi:10.1088/1742-6596/1161/1/012012

The importance of conceptualization in the use of Wolfram Mathematica 10 for a course in differential calculus

S A Montoya[1] **and D A Prada**[1,2]
[1] Grupo de Investigación en Materiales GIM, Universidad Pontificia Bolivariana, Bucaramanga, Colombia
[2] Grupo de Investigación Saber, Educación y Docencia SED, Universidad Pontificia Bolivariana, Bucaramanga, Colombia

E-mail: sergioa.montoya@upb.edu.co, duwamg.prada@upb.edu.co

Abstract. With the continuous advance in the technological tools, both Hardware and Software, a massive utilization has started in different fields of society. One of the fields in which this type of tool is used is the academic one, given the importance in the generation of spaces for discussion in teacher-student mediation, using the computer as a tool that supports the process in the classroom. In particular, the use of specialized software has made the teaching work change depending on the support in the resource, however, you cannot depend exclusively on the resource, but the teacher must be able to use the tool recognizing the benefits and limitations that the software You can submit. The objective of this work is to socialize a classroom experience developed with the students of Differential Calculus in the thematic axis of inverse functions using Wolfram Mathematica 10 software, in which it is possible to show a program limitation with respect to the Cartesian representation of certain functions inverse, which can lead to misinterpretations about their behavior

1. Introduction

In some moments of any traditional course of differential calculus it is important to realize a Cartesian model (graphic) that represents the behavior of a given function and that of its inverse function, for this it is possible to resort to a specialized Software that allows to glimpse such functions when they are not trivial or typical elementary functions. Although it is possible to know the behavior of the inverse of a function from the same function by applying a geometric criterion that consists of making a rotation of its graph 90° anti-clockwise followed by a reflection with respect to the new vertical axis, in some cases It is convenient to obtain the inverse directly. Thus, through the classroom experience that we present below, we justify the importance of conceptualizing and internalizing the most elementary notions of calculus and of course of many other areas of knowledge, also of a theoretical nature, which are currently studied in the curricula of science or engineering.

2. Basic definitions

Currently there are several academic software tools that support teaching inside and outside the classroom. In the area of natural, exact and technical sciences in mathematics, it is possible to find a wide range of programs with which exercises, simulations, images are created, etc., and in each one of the different branches of mathematics such as algebra, calculus, geometry, among others. For the subjects of basic university cycle the use of ICT tools is of great help for teacher mediation, both in hardware and software. It is the duty of the teacher to review these ICT tools in order to be sure of their optimal and appropriate use. To contextualize the experience in the use of Mathematica 10 software, it is necessary to present some elementary definitions about the functions.

Definition 1.1. A function between two non-empty sets A and B is a relation between such sets, such that every element of the output set A is related to one and only one element of the arrival set B.

The above definition is usually presented at the beginning of any course of Differential Calculus (Calculus I), however this definition has had several changes in its writing over time by example the concept of function according to Lagrange in 1797 is any expression of calculation in which such quantities enter in any way, mixed with other quantities or not with other quantities that have given and invariable values, while the quantities of the function can receive all possible values [1]. With the definition 1.1. it is intended that the student has a more formal notion of what a function is and thus avoid generating confusing ideas that usually associate such a concept with the "rule of correspondence", "equation", "formula" etc., the above can be evidenced during the didactic transposition that is generated in teacher mediation wrongly, understanding didactic transposition as a process through which the action of transposing a knowledge to a didactic site takes place [2] and not as an analogy of what is wanted to understand. Specifically, Cheval lard refers to didactic transposition to the adaptation of mathematical knowledge to transform it into knowledge to be taught [3], however this transposition produces certain effects such as simplification and delegating, creation of artifacts or production of totally new objects [4]. The contributions of the didactic transposition are to allow a didactic intervention that prevents, in the limit of the possible, the formation of inadequate or even erroneous concepts, and on the other hand allows the teacher to recognize his / her own implicit conceptions as far as the mathematics is concerned [5] and from this arises the idea of didactic engineering [6].

A mathematical object, as a function, does not exist independently of the totality of its representations, but it should not be confused with any particular representation [7]. Such ideas are expressed by students in courses after differential calculus, evidencing the lack of conceptualization in this area of mathematics. An analogous definition of function widely used based on the didactic transposition is "the variable (dependent) $\boldsymbol{y}$ it is a function of the variable (independent) $\boldsymbol{x}$ if there is a rule by which to each value of x, belonging to a certain set of numbers, corresponds a defined value of y" [8]. Another reason why some people still call a function a formula, is because in the mid-nineteenth century, mathematicians understood only the term function as an analytical expression, which had its support given that they considered a graph discontinuous corresponded no algebraic expression [8].

Another of the elementary concepts during the course is the injective function that we remember below:

Definition 1.2. Let f a function from A to B. We say that f is an injective function if and only if every element of its path is the image of a single element of its domain.

It is convenient to mention that the previous definition is commonly confused by students with the concept of function due to the uniqueness of the elements that are related. On the other hand, it is important to know the previous definition well since it allows to study the inverse concept of a function since for every injective function f it is possible to define another function from it and that is called the inverse of f and that is usually noticed by the symbol f^{-1} and that satisfies the following equalities:

$$f \circ f^{-1} = I \ y \ f^{-1} \circ f = I, \tag{1}$$

Being I the identity function. That is, the composition of the two functions results in the identity function, which means, intuitively, that by applying f followed by f^{-1} to every element the same element is obtained, in other words, it is invariant.

IOP Conf. Series: Journal of Physics: Conf. Series **1161** (2019) 012012 doi:10.1088/1742-6596/1161/1/012012

Definition 1.3. The graph of a function *f* is the locus of all related points by *f*.

3. Results and discussion

With the continuous advance in the technological tools and their easy acquisition, there is evidence of greater use of said resources in the classroom, so much that nowadays the teaching mediation is articulated with strategies supported in the TIC, such as the use of video beam, videos, social networks, etc., however the new technologies will end up eroding the curriculum and will demand a new one, articulated intimately with said technologies [9], not only with respect to the physical resource but also to the different applications that exist. The problem does not lie in the articulation of these tools, but on the contrary of the shortcomings that these may have at the programming level, which generate errors that are accepted by the learners. When a student has understood the arithmetic, algebraic and / or logical process, technological tools become a support in their cognitive process, however, if the artifact is considered as the means of solving a given problem, for example in mathematics , the unthinking use of the device can introduce distortions in the processes of teaching and learning, then the computer or the calculator does not come to demobilize the cognitive activity, but to enhance its action in new areas to reorganize the cognitive functioning of the student [9].

An integrated class gives meaning and practical significance to computer science, as it seeks the integration of knowledge in the education of the learner [10]. The experience basically consisted in using the software Mathematica 10 [11, 12] during 5 classes of differential calculus to represent any function whose inverse is not possible to express analytically explicitly [13], as is the case of the transcendental function. injective: $f(x) = e^x + sin\, x$. The result is the one shown below (Figure 1).

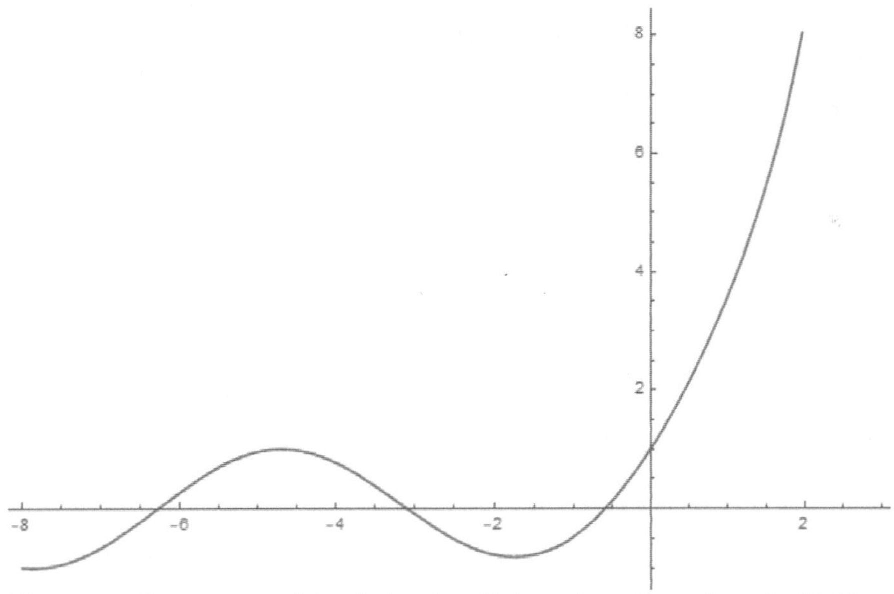

Figure 1. Cartesian model of function $f(x) = e^x + sin\, x$, through Wolfram Mathematica 10.

The response of the students as expected was accepted as they assume that the software is "advanced" and its computing power for this type of task is enough. Subsequently, the result of implementing the code to generate the inverse was shown (see Figure 2). Of course, the answer was the same, although they were surprised by the shape of the resulting curve, which was awakening a certain type of skepticism until the geometric criteria were recalled obtaining the inverse of the function from the same function and of how non-injectivity affects the definition of an inverse function.

IOP Conf. Series: Journal of Physics: Conf. Series **1161** (2019) 012012 doi:10.1088/1742-6596/1161/1/012012

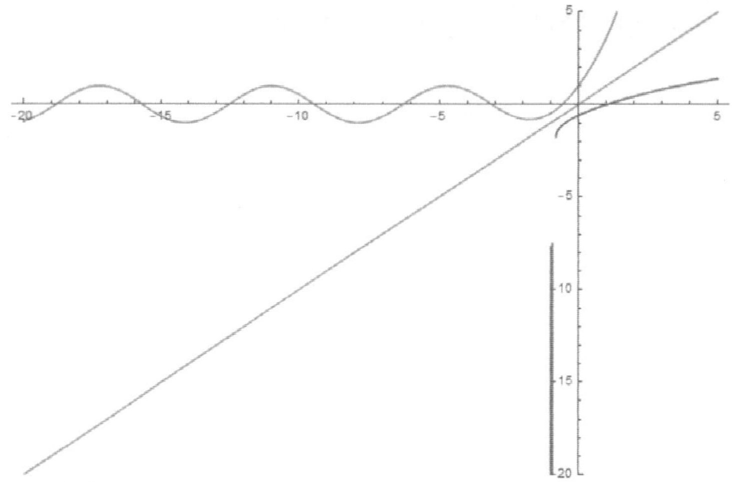

Figure 2. Cartesian model of the inverse function of $f(x) = e^x + \sin x$ green color, $I(x) = x$ orange color, f^{-1} blue color, through Wolfram Mathematica 10.

4. Conclusions

In conclusion, as in any application based on numerical methods, there is some uncertainty in the information generated and it is also subject to adequate interpretation by the user, which depends to a large extent on the background that has the concepts that support the design of the same. It is important to recognize that the use of technological tools to support the cognitive process in each person must be subject to cognitive flexibility in terms of seeing the student with his artifact, integrating the conceptual formality with the competence in the use of technological tools that the medium provides him continuously. The foregoing will demand in the mediator, a preparation not only in the theoretical part but also be digitally literate, to guarantee reflection spaces mediated by technological tools. Finally, it is convenient to propose the following questions:

Question 1. Is it possible to design an algorithm to make a representation of the inverse of a function since it is not injective in its entire domain?

Question 2. Is it possible that the use of different technological tools can generate inappropriate learning due to its limitation? How to identify this limitation? What to do in those cases?

References

[1] Youschkevich A 1917 Theory of continuous sets of points *Tohoku Math. J.* **XI-XII** pp 43-158
[2] Cantoral R and Farfán R 2004 *Desarrollo conceptual del cálculo* (Colombia: Thomson Learning)
[3] Chevallard Y 1986 La transposition didactique du savoir savant au savoir enseigné *Rev. Fran. Péd.* **76** 89-91
[4] Rumelhart G 1988 *La génétique et ses représentations dans l'enseignement Rev. Fran. Péd.* **84** 113-115.
[5] D'Amore B 2011 *Didáctica de la matemática* (Colombia: Cooperativa editorial magisterio)
[6] Artigue M 1989 Ingenierie Didactique *Rec. Didac. Math.* **9** pp 281-308
[7] Otte M 2006 Mathematical epistemology from a peircean point of view. *Ed. Stu. Math.* **61** pp 11-38
[8] Aleksandrov A D, Kolmogorov A N and Laurentiev M A 1994 *La matemática: su contenido, métodos y significado* (España: Alianza Universidad)
[9] Moreno L 2014 *Educación matemática: del signo al píxel* (Colombia Universidad Industrial de Santander)
[10] Andrade H and Gómez L 1997 *Tecnología informática en la Escuela* (Colombia: Universidad Industrial de Santander)
[11] Wellin P 2016 *Essentials of programming in mathematica* (Cambridge: Cambridge University Press)
[12] Wellin P and Calkins H 2010 *An introduction to mathematica in the classroom: version 2.0* (Illinois: Wolfram Research)
[13] González J and Martín 2014 *Int. J. Ana.* 195329

IOP Conf. Series: Journal of Physics: Conf. Series **1161** (2019) 012013 doi:10.1088/1742-6596/1161/1/012013

Student perception of the implementation of a teaching strategy based on Just in Time mediated learning and the use of information and communications technologies in the physics I laboratory course

E F Pinzón[1], **A R Lizcano**[1], **J H Martínez**[1], **G A Patiño**[1] **and D A Miranda**[1]
[1] CIMBIOS, Universidad Industrial de Santander, Bucaramanga, Colombia

E-mail: dalemir@uis.edu.co

Abstract. A Teaching strategy based on Just in Time Teaching, mediated learning techniques and the use of information and communication technologies was implemented in physics I laboratory courses to improve research skills of second-level engineering students. In this work, a study into student's perception of application of this strategy was carried out through statistical analysis of the responses given by students in perception surveys applied at the end of each semester. The study consisted of two parts: a quantitative analysis of closed test answers (questions with predefined answers) and a qualitative analysis of open answers (answers written by students). Programming codes in Python language were developed to perform the descriptive analysis of student responses from the first academic semester of 2016 (2016/I) to the first academic semester of 2018 (2018/I). Student percentage who answered "yes, they are enough" to the question if tools available in the Moodle virtual learning platform are useful for verification of the physical concepts, presented a tendency to rise (65% to 80%) along academic semesters. Also, qualitative analysis of the responses showed the need to train students in use of new technological tools (data collection software and devices) implemented in the laboratory projects in the last semesters. The above was concluded from the increased percentage of students who expressed difficulties in handling laboratory tools and devices (3.0% in 2016/I to 29% in 2018/I).

1. Introduction

In recent years, economic and environmental challenges worldwide have demanded the restructuring of higher education politics. Modern societal dynamics require abandonment of traditional teaching methods, in which students passively participate in their learning process, and implementation of teaching methods where students are actively learning [1]. Active learning comprises a set of teaching techniques that involve students in the classroom [2]. Recently, active learning implementation in courses of undergraduate programs in science and engineering has involved the integration of information and communication technologies (ICT). ICTs as Virtual learning platforms not only allow teachers to publish class activities but also provide quantifiable data about the technique effectiveness and student´s perception of it. [3,4]. Just in Time Teaching (JiTT) of active learning is a teaching strategy that allows assessment of the student's understanding of a topic before the classroom session [5]. With JiTT students answer questions proposed by the teacher according to readings and work material previously published in the virtual learning platform (Warm-Up). Then, the teacher evaluates the understanding of the key concepts detecting learning gaps and previous knowledge. This provides teachers with the opportunity to adapt class session in mediated learning activities that allow the understanding of concepts [6]. Mediated learning activities involve student-teacher interaction to enrich the student's learning experience [7,8].

ExperTIC is an innovation project developed at the Universidad Industrial de Santander within the framework of the student permanence politics of the National Ministry of Education (MEN) [9] and the Academic Excellence System (SEA) of the university [10]. This project seeks the improvement of teaching-learning experiences in the courses of the Faculty of Sciences through implementation of mediated learning and active learning (JiTT) techniques supported by Information and Communication Technologies (ICTs) [11]. ExperTIC project implementation in the Physics I laboratory courses began in the second semester of 2015. At the end of each semester, a student perception survey was applied. Perception surveys consisted of questions about aspects of implementation, like the tools available in the Moodle learning platform, teacher accompaniment of practical sessions, strategy methodology and the ExperTIC experience in general. In this work, a study into student perception of the ExperTIC strategy was carried out through statistical analysis of the responses given by physics I laboratory students in perception surveys. The results of this study became useful tools for project researchers in the decision-making process to improve the project methodology.

2. Methodology

2.1. Strategy implementation

Physics I course for engineering students has a theoretical component and a practical component. Programs and independent professors handle each component. Physics I laboratory is the practical component of this course and corresponds to 20% of its final grade. Physics I laboratory is graded as shown in Table 1. ExperTIC methodology in physics I laboratory consists of three stages: before, during and after experimental session.

Table 1. Physics laboratories grading.

Component	Percentage
Laboratory Preparation Questionnaires (CPL)	30%
Laboratory Report	40%
Final project	30%

2.2. Before experimental session

To prepare the experimental session, students reviewed the material available in the Moodle virtual learning platform, which includes research projects (laboratory guides), videos on the experimental methodology to be carried out, and data sheet models. JiTT strategy is applied to evaluate the student's understanding of the research project and to detect misconceptions before the experimental session. A set of laboratory preparation questionnaires (CPL) is proposed to students. CPLs consist of closed questions (predefined response) and open questions (student-written response) about the research project objectives, experimental methodology, data collection and data analysis. Bank of questions is reviewed and fed in each academic semester by the course teachers. Students responded the CPLs on the Moodle platform (tic.uis.edu.co) 36 hours before experimental session according to JiTT strategy. The web platform automatically closed and graded closed-questions while teachers reviewed open questions. Moodle platform configuration allows students not only to see their CPL grades but also to observe the comments and corrections performed by the teacher to each answer. The above is considered the first student-teacher feedback. In addition, students can perform a second student-student feedback in spaces outside of the experimental session.

2.2.1. During experimental session. Considering the CPL responses and misconceptions detected, the teacher began the experimental session with a third student-teacher feedback on the physical and methodological concepts not properly understood by students. Then, groups of three students carried out the experiments based on the previous revision of the published material.

IOP Conf. Series: Journal of Physics: Conf. Series **1161** (2019) 012013 doi:10.1088/1742-6596/1161/1/012013

2.2.2. After experimental session. After the experimental session, students are to elaborate the research report in accordance with the parameters published on the Moodle platform. A member of each group uploads the research report to the platform to be reviewed by the teacher. Before the next experimental session, the teacher publishes the report grade with its respective feedback for each group. Feedback allows students to perform the corrections suggested in the following report. As the last learning activity of the course, students propose and develop a final project based on a bibliographic review and the scientific method. Parameters of this activity are published on the Moodle platform. Each laboratory group performs a general presentation of the developed project and a written report.

2.3. Data analysis

Study of the student's perception of ExperTIC strategy implementation in the physics I laboratory courses was carried out through statistical analysis of students' responses in the perception surveys applied at the end of each academic semester. Programming codes in Python language were developed using Panda's library to perform the descriptive analysis of student responses collected from the first semester of 2016 to the first semester of 2018. The study consisted of two parts: a quantitative analysis of closed questions (questions with predefined responses) and a qualitative analysis of open questions (student-written answers). Number of both students who participated in the study, and students enrolled in the course are presented in Table 2 for each academic semester.

Table 2. Student population.

	2015-II	2016-I	2016-II	2017-I	2017-II	2018-I
Participating student	595	352	183	208	363	103
Enrolled student	647	400	546	501	523	516

2.3.1. Quantitative analysis. A quantitative analysis was carried out for three closed-questions. Question statements and their response options are showed in Table 3. Programming code developed allowed determination of students' percentage for each response option.

Table 3. Closed-questions and their response options.

Question statements	Response options
1. Did the tools available in the Moodle learning platform (project videos and research projects) facilitate the experimental verification of theoretical concepts?	1. Yes, they did. 2. No, they did not.
2. How often did your teacher feedback the evaluations of the research reports?	1. Every fifteen days. 2. Every thirteen days. 3. Twice per semester. 4. At the end of the semester. 5. Never
3. Did your teacher guide you appropriately during the experimental sessions?	1. Yes, he did. 2. No, he did not. 3. His guide was unsatisfactory.

2.3.2. Qualitative analysis. Qualitative analysis of open questions was performed through categorization process. For this study, two open questions were analyzed (Table 4). The responses were read carefully and organized into different categories. Each category was assigned an identification code (positive number). Response percentage for each category was determined by Python Code.

Table 4. Open questions.

1. What difficulties did you find in the course dynamics?
2. What elements would you add to the Moodle platform to improve the learning experience in the course?

3. Results and discussion

3.1. Quantitative analysis

Bar graph shown in Figure1 indicates that student perception of the usefulness of the Moodle learning virtual platform tools in the experimental verification of physical concepts tended to increase through academic semesters. Student percentage who answered "Yes, they did" presented an increase of about 15% from 2015-II to 2018-I (65% in 2015-II and 80% in 2018-I). The latter can be associated with the continuous improvement undergone by the tools of the Moodle platform, such as better-quality videos on experimental methodology, easy access to tools, improved research projects, grades history and research report guide.

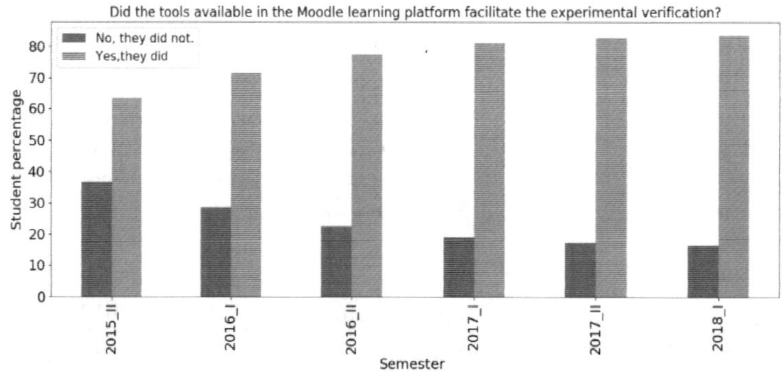

Figure 1. Bar graph for closed question 1.

On the frequency with which teachers performed feedback on research reports, the bar diagram did not show a tendency in student response behavior (Figure 2). In accordance with the ExperTIC project guidelines, research reports feedback should be carried out in each experimental session (every fifteen days) so that students will apply the corrections in the next report. The response option “Every fifteen days” had the highest response percentage in each semester but a variable behavior concerning academic semesters shows low assimilation of this practice by course teachers. Therefore, it is necessary to implement improvement actions with respect to this practice, since without appropriate feedback, it is not possible to correct errors in the research reports writing, data analysis and experimental results presentation. Since the research report is uploaded to the Moodle platform, a tool will be enabled with which the teachers can share the report feedback and corrections with each student. In the teacher training activities, teachers will be motivated to evaluate the research reports every fifteen days and make a general feedback at the beginning of the experimental session. ExperTIC project researchers every week will review the Moodle platform to follow up on the research report evaluation and give recognition to the responsible teachers with this activity.

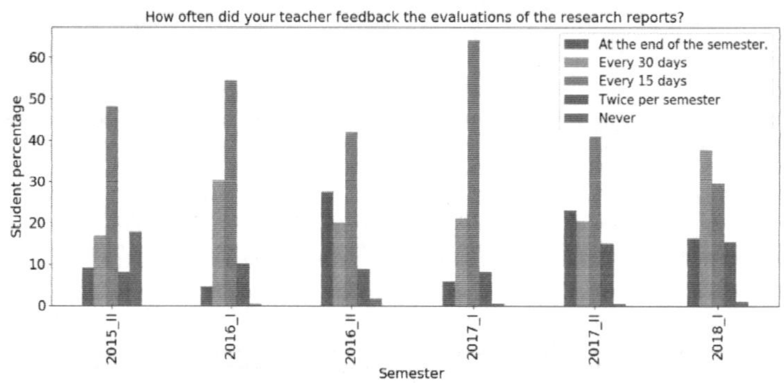

Figure 2. Bar graph for closed question 2.

Figure 3 shows the bar graph obtained from the student's responses on teacher guidance during practical sessions. The percentage of student who thought that the teacher guide was adequate showed a tendency to increase from 2015-II (60%) to 2016-II (74%). Then, it reached a stable behavior at a value of about 80%. It is worth noting that the response option "Yes, he did" presented the highest percentage value in all academic semesters in which the study was performed. The above can be associated with the active participation of the course teachers in the teacher training activities organized by the ExperTIC project researchers. In these activities, the teachers receive training on the ExperTIC teaching methodology and they can propose improvement actions based on their experience.

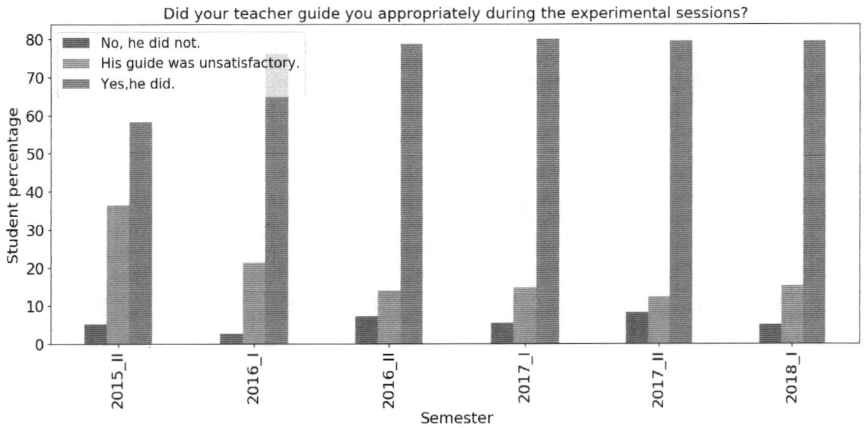

Figure 3. Bar graph for closed question 3.

3.2. Qualitative analysis

In this analysis, the responses to open questions were classified into categories through the careful reading of each of them. These categories correspond to different elements of the ExperTIC methodology implementation in the physics I laboratory courses. For the first question, 24 categories were defined, while for the second question 20 categories (Table 4) were. Each category was assigned an identification digit. Table 5 shows the eight categories with the highest response percentage for two questions. Frequency diagrams obtained for each open-question are presented below.

Table 5. Response Categories obtained in the open-questions analysis.

Code	Question 1. What difficulties did you find in the course dynamics?	Code	Question 2. What elements would you add to the Moodle platform to improve the learning experience in the course?
1	Completion of the research report	1	Quality of research project videos
2	Data analysis and treatment	2	Applications to simulate experiments
3	Neither	3	Neither
4	Experimental Set- up	4	Improvement of research projects
5	Lack of theoretical concepts	5	Information on the use of laboratory equipment and instruments
6	Bad-quality videos	6	A student chat
7	Use of laboratory equipment and instruments.	7	Material of physical concepts
8	Data collection process	8	CPL questions about research projects

Four categories related to difficulties found in the course dynamics showed a tendency to increase in the response percentage along academic semesters. These categories were: data analysis and treatment (11.0% in 2016-I to 16.8% in 2018-I), experimental set-up (6.0% in 2016-I to 12.0% in 2018-I), use of laboratory equipment and instruments (1.3% in 2016-I to 10.5% in 2018-I) and data collection (3.0% in

2016-I to 29.0% in 2018-I). According to the student's responses, response percentage behavior in these categories was associated with the implementation of research projects that involve the use of new equipment and software in the last semesters (2017-I to 2018-I). The aspects mentioned by the students can be summarized as follows: the data collection process using tablets is confusing and complicated, since tables must be restarted repeatedly during the experiment, causing time loss; software handling used in the data collection is unknown to the teacher, and air rail calibration is complicated. These aspects suggest that although the students of this era were born surrounded by technology, they do not know how to use it, so in courses they must receive training on the use of the technological devices implemented [12].

It is worth noting that the category with name "Neither" (1.3), which corresponds to the student's percentage who mentioned not experiencing difficulties in the course dynamics (Figure 4), also showed a tendency to increase along academic semesters (6.5% in 2016-I to 17.9 in 2018-I). On the other hand, three categories presented a tendency to decrease in the response percentage: completion of research report (25.3% in 2016-I to 5.7% in 2018-I), lack of theoretical concepts (3.1% in 2016-I to 0.5% in 2018-I), and bad-quality videos (2.8% in 2016-I to 0.2% in 2018-I). The evident decrease in the student's percentage who mentioned having difficulties with the research report completion was due to the creation of a general guide. This guide contains a detailed description of each of the research report parts and it is published in the Moodle virtual learning platform for all students (Figure 4). On the lack of physical concepts, students expressed that the theoretical section in the research projects is complete and facilitates experimental verification and data analysis.

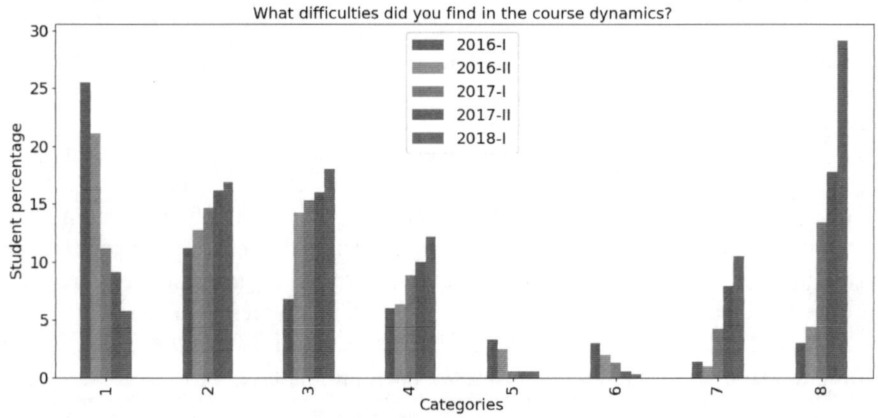

Figure 4. Frequency diagram for the first open-question.

Figure 5 shows the response frequency diagram obtained for the second open question, which referred to elements that would be included in the Moodle platform to improve the learning experience. Category with name "Neither" (2.3) that corresponds to the percentage of students who thought that the Moodle platform is efficient, and needs no more tools, presented the highest response percentage values in each academic semester. This category also showed a tendency to increase concerning the academic semesters, with 9% increase from 2016-II to 2018-I. The latter reflects the positive student perception of the continuous improvement of tools available in the Moodle platform. In addition, four more categories presented a tendency to increase the response percentage in the same academic period (2016-II to 2018-I): applications to simulate experiments (6.7% to 12.5%), information on the use of laboratory equipment and instruments (5.1% to 21.2%), student chat to solve doubts and questions (1.6% to 8.3%) and material of physical concepts (3.8% to 8.3%).

The noticeable response percentage increase for category five (2.5) can be associated with the difficulties found in the data collection process using tablets. Students thought that reviewing material (manuals or videos) on the use of these devices would save time in experimental sessions. Also, for this question three categories tended to decrease in the response percentage along academic semesters.

Category one (2.1) on improving video quality presented a decrease in 11.3%, resulting from the investment made in video recording of better-quality image, audio and information. Course teachers were trained to participate in the recording of these videos. The improvement of the research projects was observed in the reduction of the response percentage for category four (2.4). Students express that current research projects present a better composition and clearer explanation of the methodology.

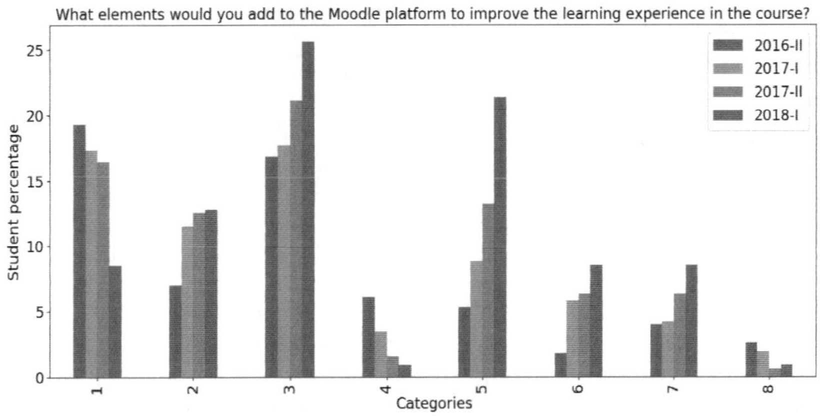

Figure 5. Frequency diagram for the second open question.

4. Conclusion

The perception study described in this paper allowed us not only to detect those ExperTIC methodology elements susceptible of improvement, but also to measure the effect on student perception of actions previously implemented, as Moodle platform tools improvement and teacher training activities conducted each semester. The next stage in this investigation is the elaboration of the general improvement plan, considering the results of this study. Among the main actions proposed are training sessions for students and teachers on the use of tablets and software used in data collection, and production of videos on laboratory equipment operation.

References

[1] McParland M, Noble L M, Livingston G 2004 *Med. Edu.* **38** 859-67
[2] Prince M 2001 *J. Eng. Educ.* **93(3)** 223-31
[3] Woods D R 2014 *Ind. Eng. Chem. Res.* **53(13)** 5337-54
[4] Deslauriers L, Schelew E, Wieman C 2011 *Science* **332** 862-886
[5] Driscoll M 2002 *Sci Res* **332** 862-864
[6] Simkins S, Maier H M 2004 *Soc Sci Comput. Rev.* **22** 444-56
[7] Hakkarainen K 2009 *Com. S. Coll. Learn.* **4** 213-231
[8] Torres V, Fanny S 2005 *Teach. Prof. Dev.* **6 (1)** 177-185
[9] Trujillo F 2011 *Enfoque de competencias en la educación: del conocimiento al uso y apropiación. Gestión del conocimiento* (Ministerio Nacional de Educación Nacional, República de Colombia)
[10] Universidad Industrial de Santander *Autoevaluación Institucional* Consulted on: https://www.uis.edu.co/webUIS/es/administracion/rectoria/acreditacionInstitucional/autoevaluacion2013.html
[11] Miranda D A, Sanchez M J and Forero O M 2017 *J. Phys.: Conf. Ser.* **850** 012015
[12] Válek J, Sládek P 2012 *Procd. Soc. Behv.* **69** 1866-1870

IOP Conf. Series: Journal of Physics: Conf. Series **1161** (2019) 012014 doi:10.1088/1742-6596/1161/1/012014

Strategy analysis Project Based Learning implementing teachers in training students in the course discipline quality management program technology in Bachelor of Pharmacy, Industrial University of Santander

L Rincón-Forero[1] and NA Rico-Bautista[1]
[1] Tecnología en Regencia de Farmacia, Universidad Industrial de Santander, Bucaramanga, Colombia

E-mail: lrinconf@uis.edu.co, ing.nolfer.rico@gmail.com

Abstract. In the UIS, the teaching-learning process is stimulated so that it has a solution in the context of existing situations and real daily life. Through interaction and reflection activities, spaces for research, analysis and critical thinking are constructed. An integral formation is developed in the student strengthening his professional field through a significant learning. The project-based learning strategy (PBA) implemented was articulated with the competency model of the technology training program in pharmacy regency. The entire execution of the process is described taking the established technical, pedagogical and technological planning as a reference. The results obtained are presented with their respective analyses applying ICTs.

1. Introduction

PBA Methodology has been a method that is already applied in several universities for quite a while, there are different experiences [1-3] that have developed around this issue forming several academic processes where research involves measuring compliance objectives, taking some its characteristics like its principles among others; there are references to research products where the PBA is related to the type of evaluation according to this methodology; others where student results based on final grades by applying this methodology developed folders associated learning environment arguing the benefits of PBA [1]; there are also teaching strategies in which children are taught the subject of social network security from another perspective and another way of applying Project Based Learning [2].

The following article is the product of an experience that is being implemented and had their bases in the framework of the guidelines given for the construction of the course Quality Management given by the program coordination of the technology in affiliated pharmacy Institute Regional and Education Projection Distance IPRED [3] Industrial University of Santander.

In the present era of knowledge work, success depends on how quickly and appropriately employees make decisions in the workplace. Much of the decision-making process is guided by facts and principles, where tacit knowledge that experts have applied while taking decisions. This knowledge is crucial for successful performance of any task. Therefore, as essential to the progress of any society always based education will play an important and fundamental role; to teach, train and educate future professionals

Program for teachers is very important to create spaces for reflection so that students can actively promote their participation in all the different aspects of knowledge, know-how and being [4]; in an organized throughout the teaching-learning process, not only in the context that it has to teach but how to how to teach; providing with this develop in student's different skills to short- and long-term continue their learning process. It is for this reason that carried out the implementation of a teaching strategy where many more resources, activities and means by which the student internalizes academic process management and assimilation of knowledge are used.

In this experience he has a phase methodology where all the elements that are to be developed are established, who are involved, the time it develops support tools for execution among other factors; Another phase is an evaluation for results of a preliminary analysis of this research and finally to generate some conclusions of the study and research questions that guided this work like a) Is the PBA can generate spaces reflection?; b) With the PBA students can gain meaningful learning?, c) The PBA strategy can be used with other strategies and learning methods?, d) With the PBA teachers provide more elements in the teaching / learning?, e) Could bring PBA students to be active part of a knowledge society [5]?, f) With the PBA facilitate an ongoing process of updating and training of teachers?

2. Methodology

2.1. Research strategy implementation

To present as would be carried out academic activities, it began with the construction of self-learning script being this the chart the course; which contains all technological, educational and technical indications to be met by the course in question; In this construction the search for the technical and information technology was necessary since the three references pedagogy, all this closely related to the competency-based training approach thus determined by the curricular structure of the program [3]. As for the technological requirements for the development of the course it has the Moodle [6] platform.

The implementation of the various activities of this strategy makes students feel more committed and motivate them to be more involved, stimulating responsibility for their learning not only from an academic perspective but from the practical and labor, discovering and strengthening of this so technical skills through the interaction of real situations; It is why Moodle has available resources [7] to perform each of these activities without losing the training intent is intended by the PBA as seen in Figure 1.

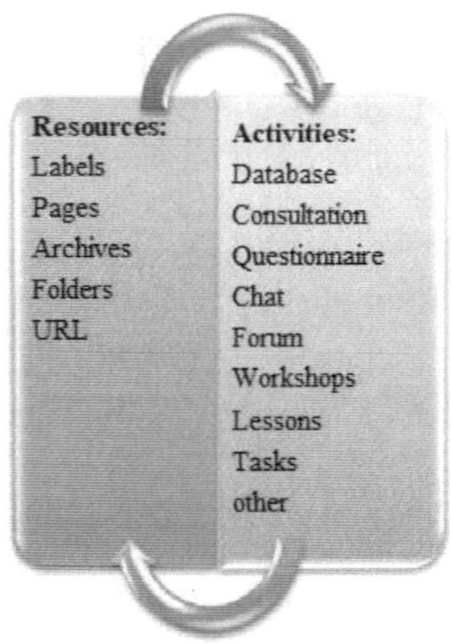

Figure 1. Moodle resources and activities.

Thus this platform through its various resources and spaces help to make an individual work through some deliverables that are known products or tasks according to the script already set, also allows the opening of group spaces from resource Forum which it generated a discussion with a number of elements and situations that encourage the construction of knowledge on a subject where each learner brings his view thus generating the involvement and participation collaboratively; in our case they were incorporated specifically Circles Interaction and Academic and Social Participation - CIPAS - as working groups of students to create opportunities for interaction and discussion of issues related to the current driving with this social interaction among them encouraging comprehensive training with the search for personal growth as an active part of the group.

Also strengthens teamwork within the group, for the particular space that each CIPA and in which its members (teachers and students) perform the feedbacks his teammates against the different contributions made around the proposed theme, from the roles each; and asynchronously evidenced how the contributions of the monitoring by the teaching building the final project is enriched.

From the technical side, it is carried out the review of the information required and subsequently selecting the same; proceeds to translate the script thematic, learning strategies, organizing students in the course and as would perform different interventions in the areas of building collaborative activities such as individual, and thus achieve the educational intentions of the course.

During Quality Management belonging to the III level Technology Program the UIS [3] which takes place this experience has key elements for the implementation of this strategy, as are the general characteristics of students and they have acquired advanced knowledge about the program, the general jurisdiction of the course involves practical involvement in the issue as it is applied to organizations where future professionals will make active and with the characteristics established from the curriculum program which It is offered in the form given distance.

The themes under which the course Quality Management develops in your large percentage handles information with a high normative component; which requires learning strategies used [8] that promotes the appropriation of knowledge and lead to a practical learning, and this tends to the student finally take decisions in the situations facing a reality, being necessary throughout their training process as well as in the professional field in which it will play well and be consistent with competency-based training [9] that was established from the training program.

And the total content of the course learning different strategies are applied; Departing from the moment of conceptualization and later appropriation and socialization of learning, it is how the student takes to from conceptualization of specific knowledge of the topic addressed both individually and in groups, in search of collaborative learning [10] in each of the stages in which the course is developed; group making available the expertise acquired since the beginning of the course, and develop skills such as designating roles, group decision making, work organization, to conclude in delivering product developed collaboratively.

Specifically, for the development of the main course work learning strategy called project-based learning [11] is used; and where harmoniously integrate this strategy achieving a significant learning. The results of this experience are very valuable and positive from the aspects such as the application of knowledge to real situations [12] to which they correspond to respond since the construction of the project, submitting written and finally lift the same, results measured by the application of evaluation rubrics to quantify the scores obtained in the two activities described above and which lead to develop the skills [13] that are required in the work context.

For the teacher also represents an enriching experience that contributes greatly to transfer all the knowledge and experience in the workplace to academic spaces training; allowing dynamic knowledge and get the maximum student academic achievement and achieve competency training intentionality proposal from the Program; as well as opening up new application scenarios of different learning techniques or strategies in other courses or signatures of the curriculum of the training program.

2.2. Strategy implementation

The Quality Management course is designed to give a general reference to the basic concepts regarding Management Systems and Regulations has established for a regency professional pharmacy in a general context. In this framework, students will learn that it is a management system, as implemented, monitoring is given and controlled, how it is the regulation and how it affects any process and relate to their professional profile.

The activities developed to comply with the process are shown below and in Figure 2:

- A review by teachers in the general context of good practice in pharmaceutical distribution facilities and process monitoring and control, licensing and accreditation issues in the course will be given.
- A problem that involves a real situation to generate the respective project is formally established
- goals that lead to student cognitive development while the respective grading rubric are established.
- The requested activities are designed, the working groups are established and guidelines for resolution are forthcoming.
- Spaces in which students must define what terms, concepts, identify "keywords" tracks and delimit the problem for this project can use the tools you have the platform as are forums and internal messaging are established.
- Students conduct brainstorming to go organizing the causes that may cause the actual problem to set in the project.
- They should then identify the needs facing the problem of the project, establishing the relationship of the objective of the activity with the themes and objectives of the course.
- Search and analysis of information through various resources that serve as support to the development of the project and thereby creating the best strategies.
- The teacher will give a brief presentation and explanation of the project giving suggestions, summary, key points and some questions that give solution routed to the project after seeing the progress of students would be generated.
- Interaction between students with significant contributions in the forum platform space where some questions you do not know on the subject arise is done.
- Feedback is performed by the teacher to each of the groups in the generated tracking to promptly clarify doubts.
- During interaction forum is generated the respective document in accordance with the guidelines established in compliance with the respective heading.
- For the solution of some item of projects complementary strategies to help strengthen the subject as in the case of audits where real context space is generated audit are used.
- Document delivery is done by a space designed exclusively for sending tasks, this shipment is made by a representative from each group.
- Feedback from the teacher is facing the product delivered.
- A session for the socialization of each product group is established and a group discussion is to clarify concepts.

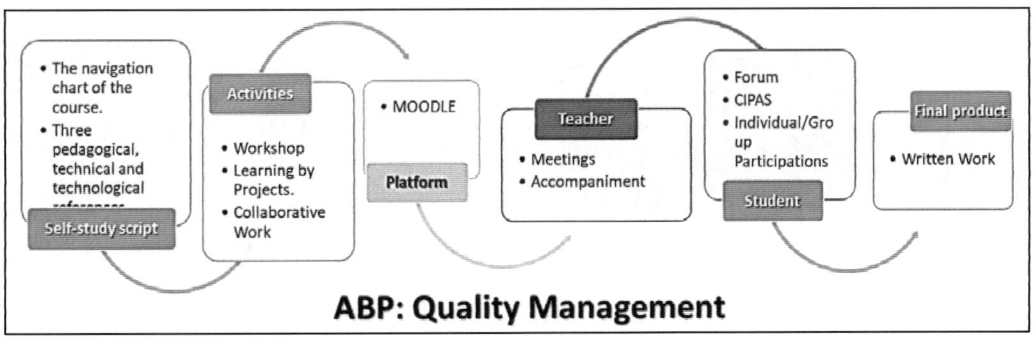

Figure 2. Strategy implementation.

3. Results and discussion

With the implementation of this learning strategy in the Quality Management course, it has been possible to identify that from the first period of 2017 to the first period of 2018 the evolution of students regarding their academic performance [14], reflected in their grades, where the behavior is evidence from inception to the current period of investigation; allowing to identify a gradual increase in the last measurement period with 91% of students with notes in a range between $4.0/5.0$ *and* $4.2/5.0$

The analysis also performed identifies the relationship between the qualification obtained by active participation in the Thematic Forum through contributions shared this space, as well as the marks obtained in the final work (includes written documents and supporting thereof) as shown in Figure 3, showing the dynamics and the correlation occurs between the scores obtained for the two activities.

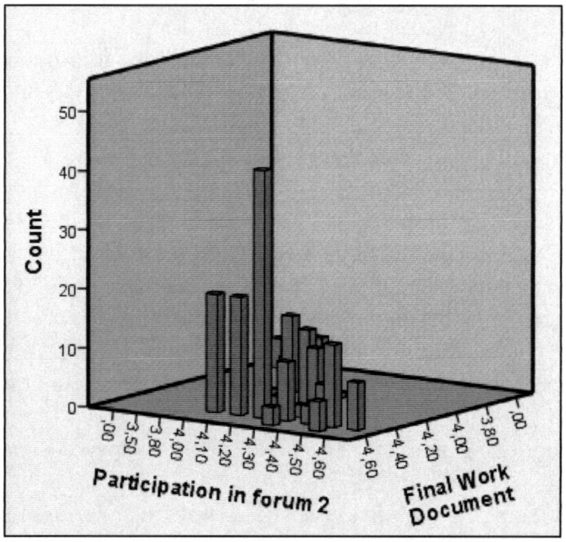

Figure 3. Tracking relationship participation in the forum against the final work.

Finally, is presented in Figure 4 the behavior of the final grade for the selected academic periods where significant growth [15] for the range 2018 *I* endnotes evidenced, which allows us to recognize how the strategy of project-based learning provides most favorable scenarios for the overall development of students [16].

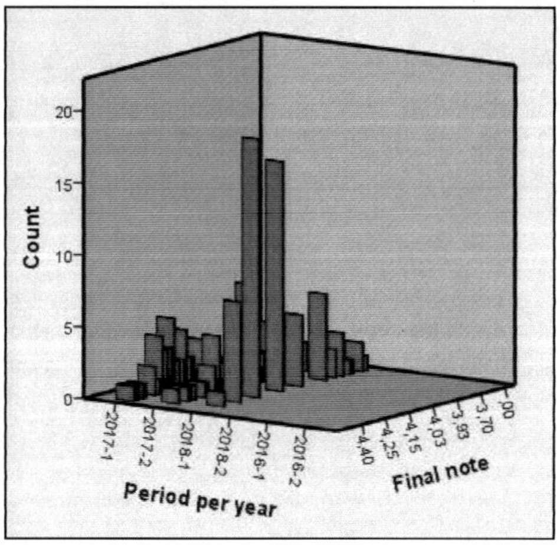

Figure 4. Final results ratio versus academic periods.

4. Conclusions

With the implementation of this experience in the institution and specifically mentioned within the course provides a great training process and interdisciplinary learning as it is useful and according to the requirements found in the workplace; carrying a feedback between academics and the real world; It is coherent with the provisions of competency-based training program and professional profile.

The behavior of the grades obtained to measure the effectiveness of the implementation of the PBA and the use of appropriate evaluation techniques, enabling the development of cognitive, behavioral skills and performance to implement a project that leads to solve a need in the actual environment.

Similarly identifies the close relationship between student achievement forum for project construction, as the qualification obtained final product comprising written work and sustaining it.

The application-based projects make the group of teachers who directs the course become actors more active in the process of teaching and learning by encouraging all students to face new experiences with new tools for learning looking strategy always a search and an optimal solution to all problems proposed. This strategy becomes a challenge for teachers; it is necessary that each knows the approaches or learning strategies, which like this can be implemented in the respective courses motivating and encouraging students to make decisions that are inherent to their professional training.

Open spaces where there is construction of collective thinking, critical analysis strengthens and generates positive changes in decision-making through joint work has in common generates products which has been active.

This experience opens the possibility to replicate in other courses of the training program, using the same way other strategies or learning techniques according to the main theme, seeking to strengthen skills for the professional future from the bimodal training offered by the Program.

References

[1] E Gregori Giralt and J L Menéndez Varela 2015 La evaluación en el Aprendizaje Basado en Problemas (ABP). Resultados de una experiencia didáctica en los estudios universitarios de Bellas Artes *Estudios Pedagógicos* **XLI(2)** 87–105.

[2] Cuesta Quintero F, Rico Bautista D, Coronel A, Portillo Ballesteros E Y, Sanchez Espinosa L C, Vera Gutierrez M D, Alvernia Acevedo S A, Rueda Rueda J S, Sanchez Torres J A, Sanchez Torres B, Bacca Picon R H and Cantillo Carreño A F 2016 *Inseguridad en las redes sociales e internet: prioridad en las escuelas de la provincia de Ocaña* (Medellín: Instituto Tecnológico Metropolitano).

[3] Tecnología en Regencia de Farmacia 2014 Reforma Curricular (Bucaramanga: Universidad Industrial de Santander).

[4] Dewar R, Leydi H and Alexci S 2017 La gamificación y arquitectura funcional: Estrategia práctica en el proceso de enseñanza/aprendizaje usando la tecnología *Revista Ingenio* **14(1)** 123-126.

[5] Krüger K 2006 El concepto de sociedad del conocimiento *Revista Bibliográfica de geografía y ciencias sociales* **11(683)** 2174298.

[6] Ros I 2008 Moodle, la plataforma para la enseñanza y organización escolar *Revista Didáctica* **2** 1–12.

[7] Nunes C S, Torres M K L, de Oliveira P C and Nakayama M K 2012 O ambiente virtual de aprendizagem Moodle: recursos para os processos de aprendizagem organizacional *Simposio Brasileiro de Informática na Educao-SBIE*.

[8] Vivas Lopez N A 2010 Estrategias de aprendizaje *Góndola, enseñanza y aprendizaje de las ciencias* **5(1)** 27–37.

[9] Tobón S 2008 *La formación basada en competencias en la educación superior* (Guadalajara: Universidad Autónoma de Guadalajara).

[10] Maldonado Perez M 2008 Aprendizaje basado en proyectos colaborativos. Una experiencia en educación superior *Laurus* **14(28)** 158-180.

[11] Martí J A, Heydrich M, Rojas M and Hernández A 2010 Aprendizaje basado en proyectos: una experiencia de innovación docente *Revista Universidad EAFIT* **46(158)** 11–21.

[12] Font V 2006 Problemas en un contexto cotidiano *Cuadernos de Pedagogía* **355** 52–54.

[13] V. G. Maura, R. María, and G. Tirados 2008 Competencias genéricas y formación profesional: un análisis desde la docencia universitaria *Revista Iberoamericana de Educación* **47** 185–209.

[14] Lucero M M 1999 Entre el trabajo colaborativo y el aprendizaje colaborativo *Revista Iberoamericana de Educación* **33(1)** 1–21.

[15] Medina Cardenas Y C, Rico Bautista D and Rosado Gomez A A 2012 El aprendizaje autónomo en los AVA *Puente Revista Científica* **6(2)** 69–73.

[16] Medina Cardenas Y C, Rico Bautista D Bautista Rico N A 2011 Calidad en la función tutorial para la gestión de entornos virtuales *Revista Educación en Ingeniería* **6(12)** 23–36.

IOP Conf. Series: Journal of Physics: Conf. Series **1161** (2019) 012015 doi:10.1088/1742-6596/1161/1/012015

Research skills development in physics laboratories located in regional headquarters of the Universidad Industrial de Santander, Colombia

E L Castellanos-Leal[1], D A Miranda Mercado[1], R F Valdivieso Bohórquez[1], J H Martínez Téllez[1], M J Sánchez Soledad[1], A R Lizcano Dallos[1], R Ospina Ospina[1] and G A Patiño Benavides[2]
[1] Escuela de Física, Universidad Industrial de Santander, Bucaramanga, Colombia
[2] Vicerrectoría Académica, Universidad Industrial de Santander, Bucaramanga, Colombia

E-mail: elecasle@uis.edu.co, dalemir@uis.edu.co

Abstract. The Universidad Industrial de Santander developed in 2015 a novel methodology adopting Just-in-Time Teaching, active learning and mediated learning strategies for classical mechanics (Physics 1), electromagnetism (Physics 2) and, waves and particles (Physics 3) laboratories, with a focus on competencies for research formation, aimed at basic cycle engineering students. In 2017, the same methodology was implemented in three Universidad Industrial de Santander regional headquarters located in Malaga, Barbosa and Socorro cities. Initially, a first visit was made to recognize materials and equipment available in each regional headquarter, for the creation of virtual learning classrooms in Moodle platform. Then, in 2018 during a second visit, students and teachers who participated in the process were interviewed and the main difficulties and opportunities that they detected in the implementation were collected, thus allowing the action plan creation that contributes to the continuous improvement in the laboratory experiences.

1. Introduction
In the natural sciences teaching, "laboratory" is the general name that is given to an activity based on observation, tests and experiments done by students, it is difficult to learn to do science or understand science, without laboratories [1]. According to Hurd [2], laboratories' purpose is to involve learners in logical use of procedures and strategies to demonstrate the scientific theories and laws implications, to provide experience to solve natural questions. Vygotsky explained the relationship meaning between language and action and how students learn in different social conditions [3]. With respect to science learning, Vygotsky's theory suggests that social interaction is crucial for learners to internalize new and/or complex knowledge. The laboratory is the place for a social exchange and ideas exploration; It is in fact a place for personal maturation and cognitive growth. Klopfer [4] emphasized that science teachers have a responsibility to help students understand scientific research nature. However, for most students, and particularly the younger ones, teachers must make a special effort to help them achieve the goal of understanding scientific research. In the past, limitations such as the shortlife equipment, the limited time of the classes for experiments and aspects related to safety and cost decreased the practical experiments effect [5]. Currently, advances in information technology have caused than computers to become a powerful cognitive tool, which extends students' investigative

skills and supports the science learning [6]. Functions provided by learning materials, shared ideas, and online simulations allow the internet to be an effective way to support scientific experiments [7].

At Universidad Industrial de Santander regional headquarters (UIS-RHQs) have recently incorporated a series of teaching-learning strategies for Physics laboratories aimed at training for research competences in students of engineering basic cycle [8]. Basic processes in science such as observation, deduction, measurement, communication, classification and prediction are integrated, with scientific competences processes (identification of variables, construction of graphs, description of relationships between variables, data acquisition and processing, design and analysis of research, and experimentation) to carry out each stage in the problems resolution [13].

2. Methodology

The same pedagogical strategy developed by Miranda and collaborators [4] focused on active learning [14], was used for the development of research competences in students of Physics laboratory belonging to the UIS-RHQs. To carry out this implementation for the first time in Malaga, Socorro, Barbosa and Barrancabermeja, the following methodological phases were established:

2.1. Review of equipment and materials available in the UIS-RHQs' laboratories

Initially, a series of visits was made in 2017 with the purpose of doing the practices verification and assemblies used in the laboratories of Physics 1, Physics 2 and Physics 3. The material collected was selected and edited to later adapt in the virtual platform.

2.2. Adaptation of digital resources for each UIS-RHQs in the Moodle platform

In the Moodle platform, the digital content (photos and videos) compiled in the previous visits was added. In addition, research projects and worksheets that are necessary for the development of laboratory practices were created. Finally, a laboratories' agenda was proposed by RHQs, where the work schedule of the laboratory groups is established during the semester. These students carry out the laboratory preparation questionnaires (CPL), reports and final projects [4].

2.3. Analysis of the satisfaction level

Two perception surveys were designed on the project, one aimed at teachers and the other at students who participated in this implementation. The first, focused on evaluating aspects of quality and usefulness of digital resources available on the platform, for the development of laboratory practices focused on research training of Physics students. The second survey for students was also designed with the intention of knowing the degree of satisfaction of the process developed in the first academic semester of 2018, in which the description of the experience in the use of virtual resources adapted to the Moodle platform is requested and its use throughout the course. Additionally, a series of visits were scheduled in each UIS-RHQs during 2018, to support teachers and students on the methodology proposed in the project and the digital resources management. The results of this initial work are detailed below.

3. Results and discussion

3.1. Perception surveys of teachers

The perception survey 2018-1 carried out with 16 physics teachers in whole university, considering that some teachers have groups in Universidad Industrial de Santander (UIS) also have in UIS-RHQs. Results are shown below.

3.1.1. Structure of support videos for research projects. Quality of videos presented in the platform to support the laboratories was associated to several parameters, that must be included in the development practice and that were previously defined. These factors are: presentation of project

IOP Conf. Series: Journal of Physics: Conf. Series **1161** (2019) 012015 doi:10.1088/1742-6596/1161/1/012015

objective, a methodology description, image and sound clarity, and usefulness that it represents for students. Perception that teachers have about this didactic resource is shown in Figure 1.

Results reveal approximately 4 points for all parameters associated with quality. Taking into account an evaluation scale between 0 and 5, it can be inferred that teachers consider that some aspects should be improved. Within the suggestions presented are: (i) improve sound quality in some videos, (ii) add support subtitles to identify names of the elements in the laboratory, (iii) give recommendations in the experiences, and (iv) make videos in group to promote teamwork among students. During first semester of 2018, new research projects were made with improved videos, where precisely the changes highlighted by the teachers were required.

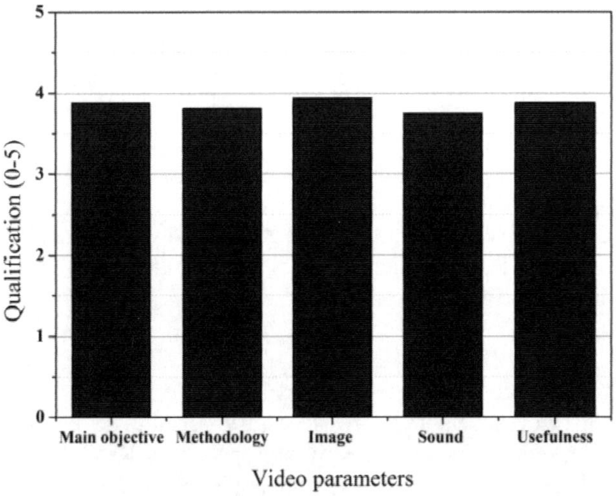

Figure 1. Evaluation of some parameters associated with quality of videos.

3.1.2. Quality and usefulness of laboratory preparation questionnaires (LPQ). In respect of questions quality (see Figure 2), teachers recommended carrying out a review and update process, because they have found with questions that are a little confusing, that generate frustration in students and that are not oriented to demonstrate the basic knowledge of concepts that have been addressed in the theory. For this reason, some corrective actions have been taken to improve questions for second academic semester, collecting each questionnaire from questions banks created in Moodle platform, and supplying them to a teacher group trained in each physics areas, who modified some questions and proposed new ones. Regarding the LPQ usefulness, teachers surveyed mentioned that open questions allowed identifying the students' shortcomings.

3.1.3. Relationship laboratory-theory and development of research skills. Relationship between laboratory and theory should be considered important in physics study and any theoretical and practical subject that involves the research skills development [8,9,10], for this reason teachers were consulted about what so related are these. Although 62.5% of teachers surveyed say that theory with laboratory are complemented very well. They mentioned that theoretical content prior to laboratory has not been seen sometimes, and students attend class very dispersed. But in other cases, where the practice is related to the topics seen in the theory, just by explaining some basic concepts, student reinforces what he has already seen and responds very well during the development and analysis of the experience [11,12].

With regard to training for research, almost 70% say that the methodology is aimed at strengthening the research skills of students. A 18% indicated that in part it does contribute, but many research competencies still need to be stimulated, only 6% affirm that there are competences that can be stimulated and are not treated, and, the remaining 6% consider that the strategies applied are not oriented towards improve this field. Additionally, many teachers said that this strategy allows

IOP Conf. Series: Journal of Physics: Conf. Series **1161** (2019) 012015 doi:10.1088/1742-6596/1161/1/012015

generating greater discipline in students and they follow a protocol for reporting results and strengthening group work. In the laboratory, students also learn to write a research report, to study a specific topic using several references, to perform a statistical analysis of experimental results, and to propose a research project in the one that set objectives and a methodology to achieve those objectives.

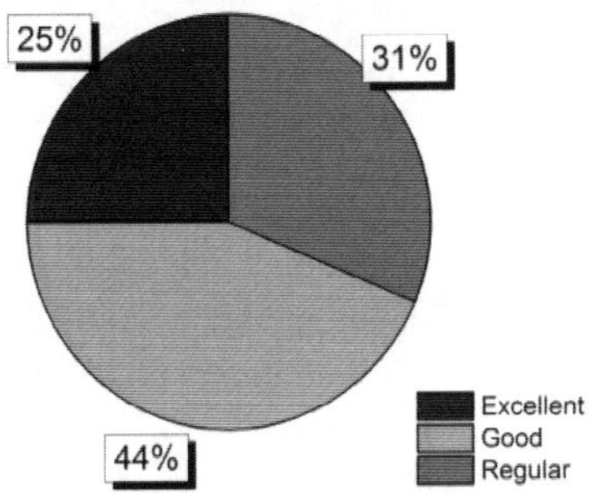

Figure 2. Evaluation of LPQ's quality.

3.2. Perception surveys of RHQs' students

In Figure 3, results of the question: Do you consider the bases offered by the project sufficient or insufficient to achieve experimental verification of theoretical concepts? carried out to 217 students (in Bucaramanga) (102 of Physics 1, 88 of Physics 2 and 27 of Physics 3) are presented. These reveal that more than 80% of Physics 1 students consider sufficient the bases given in the laboratory to verify the theoretical concepts that the subject involves, equally it happened in Physics 3 with a value close to 80%. Furthermore, in Physics 2, students didn't consider the verification of physical phenomena with the theory to be as relevant, almost 10% less than in the case of Physics 1.

Approximately 20% of surveyed in Physics 1, mentioned that they don't have problems with the laboratory practices execution. However, some difficulties were encountered with assembly and data collection, often related to the short time for practice execution. On the other hand, another group said that the greatest difficulties are in data analysis and error calculation, due to the great uncertainty generated by the laboratory elements and equipment.

Figure 4 shows the response of: Which of the following parameters caused difficulties in making the laboratory report? in the group of 102 Physics 1 students, it is mainly stated that more than 50% experienced difficulties in the handling of error obtained from experimental data compared with theoretical data and in subsequent results analysis, in addition 50% consider that graphs elaboration is also a critical factor that must be explained a bit more by teacher or within material provided on the platform. Physical principles understanding and reports writing present the lowest percentage of difficulty, 9 and 22% respectively, while conclusions construction and presentation of results are at an intermediate difficulty level (28 and 32% respectively).

As in Physics 1, a percentage close to 22% of Physics 2 students mentioned not having difficulties in performing laboratory practices, the rest of students indicate that they have presented problems in the assemblies' preparation, data acquisition and analysis of results. Physics 2 students stated that the greatest difficulty was presented in error management, data analysis and graphs elaboration. On the other hand, in the case of Physics 3, students affirm that the most common mistakes are made in the assembly and data acquisition, some assemblies are very complex and results obtained are difficult to interpret, so it is necessary to spend more time explaining methodology and equipment calibration.

In general, students recommend improve the videos quality presented on the platform for a better understanding of what should be developed in the laboratory, including programs to graph the results along with some tips to avoid making mistakes when making laboratory reports. Students consider that resources offered are good, however, in some cases the videos don't explain the process very well, and they would like to include videos or instructions on the operation of each element that will be used in the laboratory.

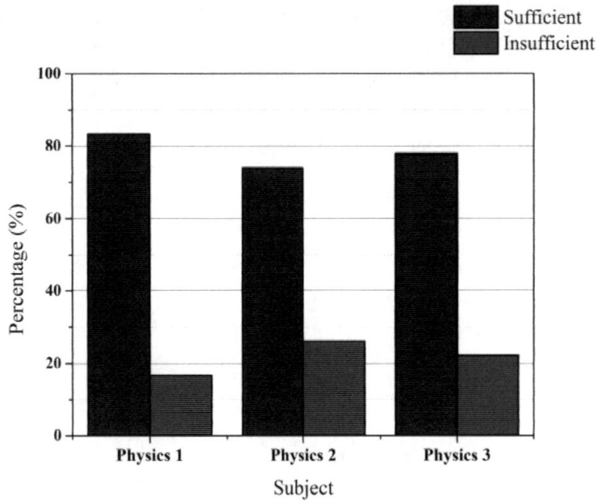

Figure 3. Answer to the question: Do you consider the bases offered by the project sufficient or insufficient to achieve experimental verification of theoretical concepts?

3.3. Implementation of the methodology in the UIS-RHQs

There are some factors that made it difficult to start the implementation of consolidated methodology. In the first place, the same laboratory equipment was not available and, the number of elements required aren't enough. In this year, two of the four regional headquarters (Malaga and Barrancabermeja) already have the same laboratory equipment as in the central headquarters, which makes it easier to articulate the research projects carried out by the students.

However, Socorro and Barbosa still don't have the same elements or assemblies. An "in situ" visit was made, and laboratory practices were verified with the intention of adapting the didactic contents (laboratory guides, videos and photographs) on the platform and thus starting with the implementation of the methodology in the proposed work schedule for the subjects. During the visit to Malaga, some inconveniences were identified in the laboratories logistics, in spite of having new equipment identical to those used at the headquarters. Table 1 summarizes the problems encountered with the group of teachers and the possible solutions proposed to each one.

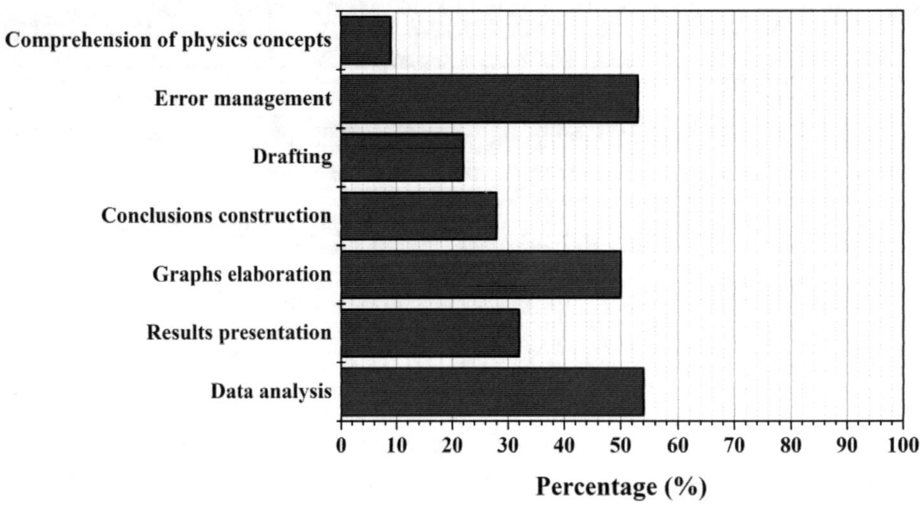

Figure 4. Parameters the generated difficulties in the laboratory report realization for Physics 1.

Table 1. Problems and solutions identified in the Physics laboratories in Malaga.

Problems	Solutions
a. The practices are not established by elements.	a. Hire 2 or 3 assistants to help with the assemblies.
b. There are assemblies that are made in the same laboratory and require a lot of care and logistics to carry them out.	b. Define mounts and responsible teacher.
c. Laboratory sessions very close that don't allow enough time to put together some complex assemblies.	c. Define class times before the start of the semester that consider the preparation of each session.
d. Lack of better organization and storage of equipment.	d. Define a storage strategy by headquarters, before and after each part of the assembly (reference and photo) and create a laboratory role for the training in "School of Trainers".
e. Missing recommendations on the use of equipment.	e. Include recommendations, limits restrictions on the use of laboratory elements in the worksheets.

Problems identified in UIS-RHQs, were addressed one by one, from the logistics of laboratories until digital resources improvement (videos and assemblies' photographs), with the purpose of being passed and delivering an improved version in the following semester.

3.4. Perception of the implementation by students of the Regional headquarters

Although the number of students surveyed in the UIS-RHQs were lower than Bucaramanga, many similarities could be found in the project perception. The students also suggest that more videos with higher quality should be added for the laboratory development, where objectives and methodology of the research project were more explained clearly, they also insist that more laboratory elements are required.

4. Conclusions

To improve the coordination of each practice developed in the laboratory, we have been working together with the theory and laboratory teachers in a teacher improvement program called "School of trainers", focused on providing teachers with the aspects pedagogical, didactic and practical in the development of the subjects intervened with the proposed methodology. In this space, the pedagogical concepts application linked with the development of the laboratories is analyzed critically: active learning, Just-in-Time Teaching (JiTT) and the mediated learning. It has also been possible to recognize quality parameters for audiovisual resources development and identify the processes necessary for the organization of evaluation instruments, in the context of the proposal.

In the physics laboratories, some research training activities are carried that include the projects execution. These research projects extend the students' perspective because allow them to strengthen research competences based on active and mediated learning strategies. An increase of students' autonomy, greater command of concepts, tools and methodologies required to carry out the experiences, and reports of the executed projects preparation were observed; the reports are made in scientific journal formats.

Acknowledgments

We are grateful to Academic Vice-rector and Physics School UIS for support the ExperTIC project, teachers and students of the laboratories that participate in this wonderful experience.

References

[1] Trumper R 2003 *Science and Education* **12** 645
[2] Hurd P D 1969 *New Directions in Teaching Secondary School Science* (Chicago: Rand McNally)
[3] Vygotsky L S 1978 *Mind in Society* (Cambridge: Harvard University Press)
[4] Klopfer E L 1990 *Learning Scientific Enquiry in the Student Laboratory* Hegarty-Hazel E(ed) (London: Routledge)

[5] Kun Yuan and Jian Sheng 2007 *J Sci Educ Technol* **16** 451
[6] Chia-Yu Liu et al 2017 *Computers and education* **105** 44
[7] Sufen Chen et al 2012 *Physical review special topics Physics education research* **8** 020113
[8] Miranda D A, Snchez M J and Forero O M 2017 J. *Phys Conf. Ser.* **850** 012015
[9] Hodosyova M, Utla J, Vanyova M, Vnukova P and Lapitkova V 2015 *Procedia - Social and Behavioral Sciences* **186** 982
[10] Murat A, Gokce A, Huseyn M P and Mehmet A O 2016 *Computers in Human Behavior* **57** 334
[11] Jingyin W, Donghui G and Min J 2015 *Computers in Human Behavior* **49** 658
[12] Wilcox B R and Lewandowski H J 2017 *Physical review special topics. Physics education research* **13** 010108
[13] Tristancho J A, Contreras L E and Vargas L F 2014 *Educación y desarrollo social* **8** 28
[14] Novak G M, Patterson E T, Gavrin D and Christian W 1999 *Just-in-Time Teaching: Blending Active Learning with Web Technology* (New Jersey: Prentice Hall)

IOP Conf. Series: Journal of Physics: Conf. Series **1161** (2019) 012016 doi:10.1088/1742-6596/1161/1/012016

Implementation of Just in Time Teaching methodology in mathematics and natural sciences study in high school institutions of Colombia northeast

E L Castellanos-Leal[1], **D A Miranda Mercado**[1], **J H Martínez Téllez**[1], **R F Valdivieso Bohórquez**[1] **and E J Vera Mercado**[1]
[1] Escuela de Física, Universidad Industrial de Santander, Bucaramanga, Colombia.

E-mail: elecasle@uis.edu.co, dalemir@uis.edu.co

Abstract. The notably undergraduate student desertion in higher education is considered one of the main difficulties in worldwide education systems. The main factors associated with this phenomenon in Colombia are related to low academic entrance skills, economic problems of students and aspects related to socio-occupational orientation and adaptation to the university environment. At Universidad Industrial de Santander an initiative has emerged since the beginning of 2017 that seeks to strengthen the entrance skills to higher education of high school students in Colombia northeast, through "Spachovsky project" that intervenes in the teaching and learning processes of three subjects: mathematics, physics and chemistry of tenth and eleventh grade, through an active learning strategy implementation: Just-in-Time Teaching whose essence is the students' previous class preparation, through questionnaires and didactic material in a web platform, which has a positive effect on what happens during the time subsequent to it with an appropriate teacher's mediation. The contents are managed through virtual tools such as: links to digital books and teaching videos through Moodle platform, which together with session preparation questionnaires, and appropriate mediation by the teacher will favor study habits and the autonomous learning of students.

1. Introduction

The World Development Report 2018: "Learning to make reality the promise of education" from the World Bank argues that without learning, education cannot be the determining factor in ending extreme poverty, generating opportunities and promoting shared prosperity. Even after attending school for several years, millions of children cannot read, write or do basic math [1].

In Colombia, MEN (Ministerio de Educación Nacional) makes efforts to provide guidelines for teaching practices at basic and secondary education that ensure that all students receive a high-quality education, which contributes to the development of basic skills for entry into higher education. It was found that the average scores are significantly lower than the average of the Organization for Economic Cooperation and Development (OECD) in PISA test, which assesses mathematics, reading and natural sciences skills. In mathematics, Colombia score (376) is lower than obtained by 61 countries and very similar to those found in the countries with the three lowest scores. In science and reading as well as in mathematics, results are also discouraging; less than 57 and 53 countries, respectively [2]. The current situation of Colombian education reveals a problem and need for improvement in the teaching and learning processes that students receive before higher education. In the light of contributing to the level of education that precedes the university, "Spachovsky" project aims to answer the research question: how can the disciplinary skills of secondary students be strengthened with the information and communication technologies (ICT) mediation and implementation of active teaching and learning strategies? Therefore, Universidad Industrial de Santander (UIS) teaching and research experiences will be used for the implementation and standardization of strategies and technologies using ICT as support [3, 4], following the MEN guidelines for student permanence and graduation management model (2005) to work in conjunction with media education institutions (IEM), with the objective of teaching and learning practices transformation and strengthening the appropriation and development of knowledge, skills and necessary attitudes for students and teachers to perform productive activities associated with scientific research.

This initiative involves a training program for teachers of mathematics, physics and chemistry about pedagogical strategies used in this intervention. This is a transcendental stage, because some different investigations call attention to the need for teacher training to respond learners needs; the continuous training for teachers is presented as a priority strategy to raise the education quality and essential axis in view of global education system improvement [5].

2. Antecedents

In order to contribute to strengthen basic competences of high school students, from the pedagogical practices transformation, it is necessary to innovate in the teaching strategies that allow to overcome and complement traditional activities characterized by the professor's exposition of knowledge that translates into students who copy from the blackboard or, from books to their notebooks without questioning [6,7]. "Spachovsky" project adopts and articulates two important components to achieve its objective as described below.

2.1. Just in Time Teaching (JiTT)

This pedagogical strategy allows to establish a teaching organization focused on eliminate the student's passivity, the transmitted knowledge memorization, using a didactic response, and orienting the classes according to learners needs. JiTT concentrates on the student while the teacher performs a facilitation process (mediated learning) not only in the processes of knowledge incorporation but also in the evaluation processes. Due to its flexibility, JiTT can be applied to a variety of tasks, courses and disciplines, and can be easily combined with other pedagogies, it is an excellent way for students to prepare the class, it can also be used by teachers to discover students' misconceptions or to provide classroom activities examples and develop disciplinary skills [8]. This strategy pursues three main objectives [9]: (i) Structure time out of class for maximum learning benefit. (ii) Maximize class session effectiveness by allowing teacher to work on specific needs detected before class. (iii) Create and maintain team spirit. In practice, to ensure the study different contents of a subject, this strategy allows teacher to know the level of understanding of a topic through questions that motivate the student to explore and be interested before taking the class, this through session preparation questionnaires (SPQ) that are intended to obtain a broad set of responses that illustrate the thinking processes of the students, while informing them of the class activities before it [3,10].

2.2. Incorporation of ICT in teaching and learning processes

"Spachovsky" project integrates technological resources such as Moodle to generate learning environments that encourage active, reflective and meaningful learning, as well as analysis and synthesis, decision making, problem solving, among others. Moodle will be the virtual environment in support of teaching and learning, in which digital teaching resources will be made available for each subject and each academic period, these resources include SPQ, readings and videos that will be carried out as part of pedagogical support that want to provide the students before entering their classes. This new environment will modify "way in which students approach learning and that varies their knowledge acquisition strategies since they initiate students in the development of new ways of working and organizing with others in which underlying a new methodological approach,

IOP Conf. Series: Journal of Physics: Conf. Series **1161** (2019) 012016 doi:10.1088/1742-6596/1161/1/012016

collaborative work" [11]. In the same way, its online and automatic using generates improvements in communication between teacher-teacher, teacher-student and student-student [12,13].

3. Methods and materials

For implementation of teaching and learning strategies mediated by ICT, a feasibility study will be previously carried out to know if the high school institutions have the characteristics to achieve the project objective. The main requirement is the provision of computer equipment with internet for teachers and each student. On the other hand, a user is required in the Moodle educational platform to access the digital resources that will be created at the start of the pilot test.

The project initially contemplates a teacher training stage in teaching and learning theories that define the methodology to be implemented. Moodle offers a set of tools that help the teacher in the monitoring and evaluation. In this way, IEM teachers will use the tools at different moments of the process, either through workshops or questionnaires that will be designed incrementally. This will allow using the evaluation from the formative perspective since its purpose is to make adjustments during the march. Different measurement and evaluation technologies for learning will also be integrated, as well as different types of questions and measurement instruments to contemplate and strengthen various learning styles, as well as Bloom's taxonomy to categorize the measurement instruments and their results to facilitate design and information analysis [14].

For project execution, four (4) phases are proposed with nine (9) stages in total that will allow the project maturation and consolidation, as shown in Figure 1. First activity consists in the material development required by the project (material for teacher training in teaching and learning strategies, ICT, Moodle and learning modules development for tenth and eleventh grade).

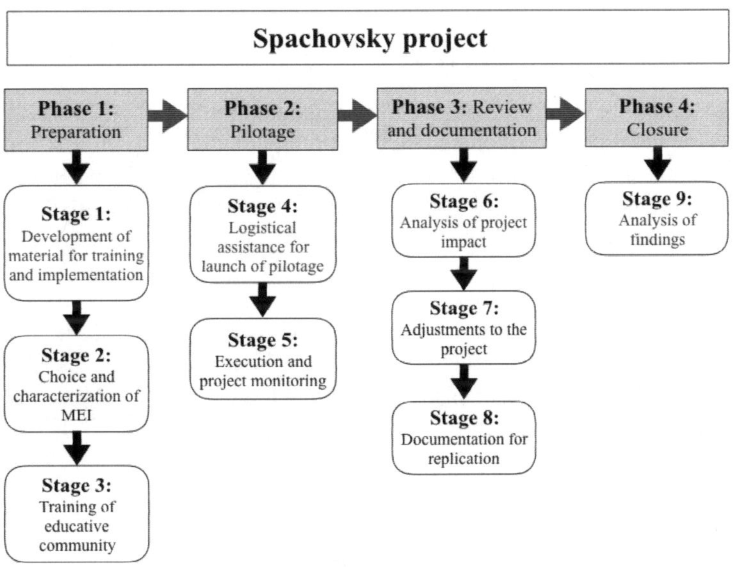

Figure 1. Phases of Spachovsky project.

Three high school institutions will be selected through non-participant observation, a characterization will be made with information that accounts for following variables: (1) pedagogical strategies used by IEM science teachers in their pedagogical practices, (2) teaching and learning resources used by teachers, (3) academic results in each discipline for two previous school years, (4) prior knowledge of the students who will participate in phase 2, (5) curriculum of each subject that will be intervened in the tenth grade, (6) science teachers profile, (7) current conditions of classroom environments and their installed capacity, (8) socio-economic stratum of students, (9) technological resources and connectivity availability outside of their institution. Teachers will be previously trained in the use of the methodological strategies that support the project.

In second phase (piloting), necessary actions will be deployed to organize material in the educational platform. Didactic material developed, and the measuring instruments will be incorporated into the virtual learning tool (Moodle) to initially intervene in the tenth grades. Following the regular academic IEM's calendar, some visits "in situ" for the accompaniment and follow-up necessary to solve possible pedagogical or technological difficulties will be necessaries.

Taking as reference the characterization carried out in phase 1 and the information gathered in the monitoring and follow-up during the project execution in different components of phase 2, in phase 3 the project impact will be determined, identifying strengths and possible aspects improvement taking into account administrators, teachers and students satisfaction in relation to the proposed mediation strategies. Subsequently, with results obtained in the previous stage, necessary adjustments will be made for the new staging and reference documents will be made on the use of the elaborated material and teaching - learning - evaluation strategies.

Finally, closure phase involves analyzing results of the implementation with tenth grade and making the assess consolidation. "Spachovsky" project covers a series of actions that are framed as a strategy of contribution to teacher training that is understood as a decisive factor for practice transformation of educators.

4. Results and discussion

This work will show results obtained in the first phase of the project (preparation), in which mediation strategies that will be used in the next phase. Virtual digital resources were created (links to books in the *UIS*'s digital library, educational videos related to the topics of interest and SPQ development).

4.1. Mediation strategies used in strengthening of skills for autonomous learning in high school students

Based on excellent results obtained in the research carried out in 2015 by Miranda and collaborators with basic engineering cycle students, using JiTT strategy, together with "School of trainers", a designated space to training of laboratory professors (Physics 1, 2 and 3) on how to implement the strategy and organize the work before and during the class [3, 14]. It was decided that for Spachovsky project a similar methodology.

4.2. Supporting didactic resources and bimedia material

Initially a bibliographic search, collection and processing contents that should be treated in grades 10 and 11 grade was carried out. Selected books were chosen from the content distribution platform for primary, secondary and higher education institutions known as Ebooks 7-24. Bibliography used for mathematics, chemistry and physics are summarized in Table 1.

Table 1. Bibliography used to study the contents of mathematics, chemistry and physics.

Subject	Support bibliography
Mathematics	Mathematics I, II, III y IV, Patricia Ibáñez Carrasco, Gerardo García Torres. Ed. 1, 2011.
	Pre-calculus, Mathematics for calculus. James Stewart, Lothar Redlin. Ed. 7, 2016.
	Introductory university mathematics, Franklin D. Demana, Bert K. Waits, Daniel Kennedy, Gregory D. Foley, Robert Blitzer. Pearson Education, Ed. 1 2009.
Chemistry	Chemistry, Raymond Chang. Ed. 11, 2013.
	Chemistry I and II, Skills approach, García Becerril María de Lourdes. Ed. 2, 2010.
Physics	Physics fundamentals Volume I, Raymond A. Serway, Chris Vuille. Ed. 9, 2012.
	Tippens, Paul E. Physics, Concepts and applications. México, 6ª.Ed., McGraw Hill, 2001.
	Physics fundamentals, Wolfgang Bauer. Ed. 1, 2014.
	Gómez Gutiérrez, Héctor Manuel y Ortega Reyes, Rafael. Physics I. Editorial CENGAGE Learning. México, 2010.

The links of guide texts are linked in Moodle platform along with explanatory videos of each topic. Videos were selected carefully, from YouTube, thanks to professors' contribution from different parts around the world (especially from Colombia and Spain) who carry very concrete and useful explanations in their academic channels. Subsequently, SPQ were structured taking into account key concepts of each topic, this formulation work began with minor complexity questions and was increasing its level from 1 (for the simplest question) to 4 for the highest level of complexity. Below are shown some physics and chemistry questions for tenth and eleventh grade, respectively.

4.3. Basic question in Physics 10th (level 1)
A box moves to the left in the negative direction of x-axis while applying a force ***F*** in four different situations. As shown in Figure 2. Which of the following is the correct order of the amount of work done by ***F*** from the most negative to the most positive?

a) 2, 1, 3, 4. ✔
b) 3, 2, 1, 4.
c) 1, 2, 3, 4.
d) 2, 3, 1, 4.

In previous example, a first level of knowledge can be evidenced, where the student is asked to remember or know the basic "work" meaning, which is related to the action of an external force that disturbs an object and produces a displacement in some direction. In this case, the option of comparing work magnitudes performed by different forces is generated.

For Chemistry, in eleventh grade students, following four questions are shown as an example of the different degrees of complexity, taking into account the four proposed levels.

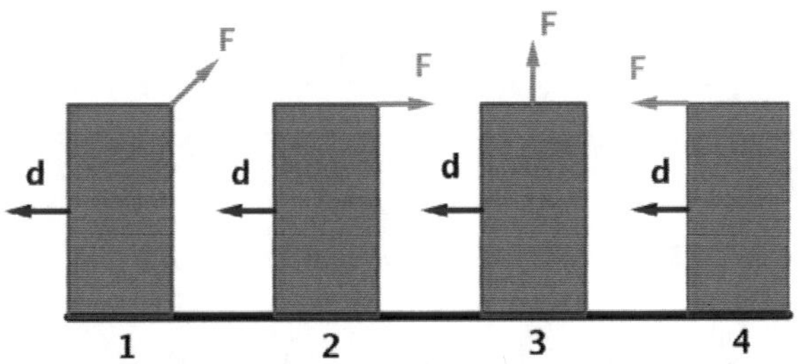

Figure 2. A constant force (F) application and a displacement (d) generated in a box.

4.3.1. Question 1 (level 1). Is each of the following statements true or false?

a) The mutual attraction exerted by the atom's nucleus on shared electrons constitutes a covalent bond. (True).
b) Elements such as beryllium and boron, with only six and four electrons respectively, are chemically stable. (False).
c) Liquid or molten covalent compounds do not conduct electricity because there are no ions present. (True).

4.3.2. Question 2 (level 2). Which of the following equations best represents the reaction shown in Figure 3?

a) $6A + 4B = 2C + D$
b) $A + 2B = 4C + 2D$

c) $3A + 2B = C + 2D$ ✔
d) $6A + 4B = 4C + 2D$

4.3.3. Question 3 (level 3). What is the bond type and stoichiometry, if A and B elements whose atomic structures are represented in Figure 4 are combined? (Electrons in red color).

a) Ionic bond and A_2B stoichiometry.
b) Covalent bond and A_2B stoichiometry.
c) Ionic bond and AB stoichiometry. ✔
d) Covalent bond and AB stoichiometry.

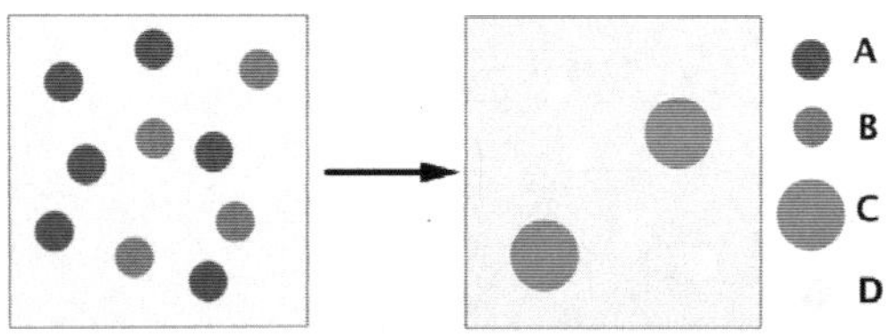

Figure 3. Chemical reaction between reagents A and B, and products C and D generated.

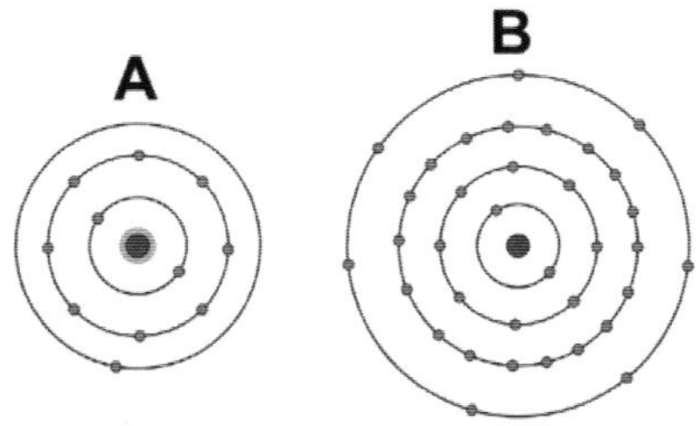

Figure 4. Graphic representation of A and B atoms.

4.3.4. Question 4 (level 4). From the gaseous reagents in *A*, the most adequate representation of the product generated taking into account the limiting reagent, within the options shown in B, C and D (see Figure 5) corresponds to:

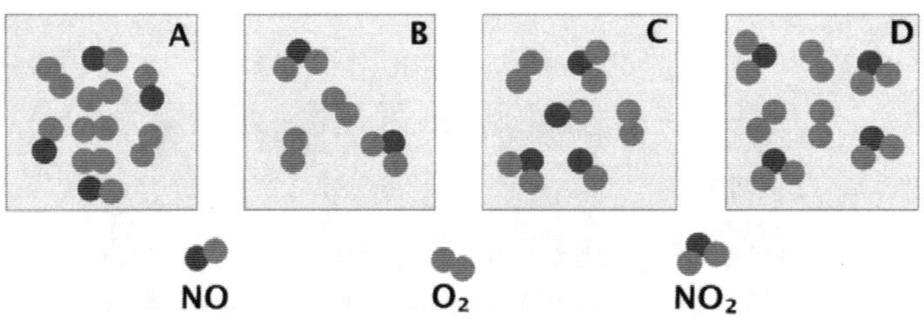

Figure 5. Graphical representation of reaction between NO and O_2 (A) and their possible combinations (B, C, D) of the products generated.

a) Option B represents the minimum amount of NO_2 that can be formed with the initial reagents.
b) Option C because the compound doesn't require more O_2 molecules to react.
c) Option D because the limit reagent is NO and it is totally transformed, leaving O_2 excess. ✔
d) None of the options, because the excess reagent is NO and not O_2.

The first question (false/true) seeks that through statements that generate doubt, students understand a fundamental concept in chemistry: the bond types. This question generates in the student the need to remember the main characteristics of bond types and the chemical nature of the elements that are grouped in the periodic table. The student must be clear about the term "chemical stability" associated with the electronic configuration of the different elements.

Questions 2, 3 and 4 show a higher level of complexity, which starts with the representation of a chemical reaction that involves four visually distinguishable elements (different color and size), two (A and B) that come together to form two new elements or substances (C and D). In this case, the student must know the law of defined or constant proportions. "When two or more elements are combined to give a certain compound, they always do so in a constant mass ratio." On the other hand, question 3 has a slightly higher level because from a basic representation of the electronic structure of two different atoms, the student is asked about the type of link that would be formed by joining the two elements and what would be the stoichiometry of the resulting compound. A student must know how the different bond types (ionic, covalent and metallic) are formed and what the stoichiometry of the resulting compound would be like. Finally, question 4 is the most complex, although it assumes control of chemical reactions and stoichiometry concepts, adds another important factor to consider in a reaction, the limit reagent and three different options that the student should analyze and compare with 4 response options.

5. Conclusions

With this project, it is expected to generate a great impact on the following educational aspects in the basic and secondary students: (i) Greater satisfaction on the part of teachers to see their expectations and learning objectives achievable. (ii) Greater satisfaction on the part of students when achieving their objectives, obtaining good grades, using their time better and improving their study habits, which can influence good results also in other subjects they are studying and even in the future subjects. And (iii) strengthening education quality in what corresponds to pedagogical training of those who teach science in secondary education.

Acknowledgments

Spachovsky project has been carried out thanks to the management of several dependencies of the UIS. La Facultad de Ciencias (2017) with the initiative of this project, Rectoría and iPred with its financial support and Escuela de Física for training and meetings provision with IEM linked to the project: Institución Educativa Llanadas-Lebrija, Colegio Integrado Mesa de Jéridas-Los Santos and Colegio Técnico Agrícola de Cáchira-Norte de Santander.

References

[1] Banco Mundial 2018, Informe sobre el desarrollo mundial 2018: Aprender para hacer realidad la promesa de la educación panorama general (Washington: Banco Mundial)
[2] ICFES 2013 *Colombia en PISA 2013 Informe nacional de resultados* resumen ejecutivo (Bogotá: ICFES)
[3] Miranda D A, Sanchez M J and Forero O M 2017 *J. Phys.: Conf. Ser.* **850** 012015
[4] Cardelli J J and Duhalde M A 2001 Formación docente en América Latina. Una perspectiva político-pedagógica *Cuadernos de Pedagogía* **308** 38-45
[5] Ministerio de Educación Nacional 1998 *Lineamientos curriculares en ciencias naturales y educación ambiental* Consulted on: URL https://www.mineducacion.gov.co/1621/article-89869.html
[6] Jua S K, Sarwanto and Sukarmin 2018 *J. Phys.: Conf. Ser.* **1022** 012027
[7] Nisa E K, Koestiari T, Habibbulloh M and Jatmiko B 2018 *J. Phys.: Conf. Ser.* **997** 012049
[8] Novak G M, Patterson E T, Gavrin D and Christian W 1999 *Just-in-Time Teaching: Blending Active Learning with Web Technology* (New Jersey: Prentice Hall)
[9] Ausubel D P 2000 *The Adquisition and Retention of Knowledge: A Cognitive View* (Netherlands: Springer)
[10] Mercado B I and Oyola M C 2012 *Revista Escenarios* **10 (1)** 17
[11] Perkins M and Pfaffman J 2006 *Sci Teacher* **73(7)** 33
[12] Chrissis M B, Konrad M and Shrum S 2011 *CMMI for development: Guidelines for process integration and product improvement* (New Jersey: Addison-Wesley)
[13] Lizcano A R, Miranda D A, Martinez J H and Sanchez M J 2017 *9th International Conference on Education and New Learning Technologies*
[14] Bloom B S, Hastings J T and Madaus G F 1977 *Evaluación del Aprendizaje* (Buenos Aires: Editorial Troquel)
[15] Benjamin S Bloom 1971 *Taxonomía de los Objetivos de la Educación* (Buenos Aires: Librería Ateneo)

IOP Conf. Series: Journal of Physics: Conf. Series **1161** (2019) 012017 doi:10.1088/1742-6596/1161/1/012017

Efitec, software for the evaluation effectiveness model of the technical

R D Merchan-Osorio[1], J Florez-Florez[1], W F Caro-Cely[1] and J A Verdugo-Acevedo[1]

[1] Grupo LERD, Fundación Universitaria Juan de Castellanos, Tunja, Colombia

E-mail: rdmerchan@jdc.edu.co

Abstract. Nowadays sports levels are increased to a high degree thanks to the use of tools and instruments that combine the use of technology with the sciences applied to sports. Football is not the exception; thanks to these technological and scientific advances it has been possible to improve not only the sports performance but also the way in which each one of the factors that determine the sport form is controlled. In Colombia today, you can optimize football through specialized software that allows you to measure, quantify, plan and organize the tasks and means of this sport. For this reason, this study aims to design an application that allows measuring the level of technical effectiveness of the shot on goal in football. This will help to improve sports performance and thus increase technical levels, allowing a better planning of the daily structures of sports training and thus contribute to the improvement of this sport by optimizing the times and the easy handling of technology, computer science and more important the results in the competition.

1. Introduction

Team sports and, more precisely, interaction sports with shared space (DIEC) [1], make up and are part of the select group of disciplines practiced with high training loads and in need of training differentiated. These processes of organization, planning and ordering of each of the components, processes, protocols and procedures need more and more sophisticated means to perform the control and optimization of the factors that determine the sports performance in these disciplines.

The high competition needs now greater and faster information of what happens at every moment of the season. The levels of technical effectiveness of the players every time plays a priority role to determine the final result of the game, there is no time to wait on the part of the coaches, managers and actors of the sport, to know the current state of the athletes; for this and many more reasons it is vital to develop software that determines in a specific way in which state the soccer players are at the moment of executing the technical action of the shot on goal since this is the most important action in the game of soccer.

This application allows the football coach or coach to know exactly the level of the athlete and have a database for monitoring and controlling the levels of technical effectiveness of the players' shooting, during the training sessions thus also allowing finding the methodological models or the specific means for the improvement of this sporting gesture.

The present work presents results obtained by this tool and the operation of this as it allows characterizing the levels of technical effectiveness of the shot to door.

IOP Conf. Series: Journal of Physics: Conf. Series **1161** (2019) 012017 doi:10.1088/1742-6596/1161/1/012017

The study involved different populations of amateur, semi-professional and professional players from different cities of Colombia who showed a lot of interest when making the tests for the novel and easy to understand.

The protocols were initiated with the setting and presentation and explanation of the tests, then piloting was carried out to discriminate margins of error and later the test was made.

After this information was analyzed, interpreted and performed the corresponding statistics to know how reliable the tests were, this was verified and corroborated with the payroll completed by experts who already knew the tests and gave their concept through trials. That they applied.

2. Experimental methodology

The methodology proposed for this study is as follows: application research according to (Padrón G 2006) [2]; the start is a problem situation, it is intervened, and a detailed description is made under criteria of scientific research methodology.

Then the theory is selected, it is exposed, and the basic and specific concepts are given. The next step is to examine the situation described above and derive the corrective actions of the prototype that was exposed in phase two that could solve the problem situation all this under systematic parameters and associated instrumentation.

Next, in the next stage, the prototype test tests that are born from the theoretical components of the previous stages are carried out and the probability that the new model or the application designed to solve the problem is determined.

For this study, it corresponds specifically to the protocol of evaluation of the level of technical effectiveness of the shot to goal which aims to measure and quantify the current state of the athletes under study. It aims to improve and control more and more the performance of athletes who practice this type of sports and improve technological processes and the use of new technologies applied and combined with science and research methodology.

Based on these theorists, an analysis, design and implementation of the system for the evaluation and control of subsequent sports gestures is carried out. At the end of all this process, the verification is carried out by experts in the combined areas for this work, so as to be able to reengineer and improve the application.

The methodological model presented in this study is of the Scrum type (Agile Projects ORG, 2014), which allows the system to be developed by means of stages, in each case part of the application is obtained, which is validated by the clients in this case the coaches and players that guarantee the proper functioning of the application.

The phases of the process are the following: information and spreadsheet, tests or test, evaluation and execution of the test, control and analysis thereof and finally verification.

In each one of them, specific activities are carried out for the development of the software and the research process.

2.1. Research

This stage the characteristics, the contents and the working functions of the work are shown, the variables of the operating system are established; an analysis is also carried out, the interpretation of the data, the methodological, technical and technological framework of each sub phase. In this phase, a document is generated. In this phase, an initial document with descriptive specifications is generated, plus detailed techniques of the investigative process.

- The purpose of this research work is to characterize the level of technical effectiveness of the players under study, by using an evaluation model that allows knowing the technical status of the established groups, more specifically the shot on target, which is the determinant variable in this sport modality.
- During the implementation process, a first test called the pretest was performed, which determined the initial state of the application. After four months, a second test called posttest

was carried out. These two applied tests are composed of the technical variable: shot on goal, chosen and scientifically based thanks to the analysis and study of previous research works [3].

2.2. Planning
In this step of the research process are defined scopes, limitations and external factors involved in each stage of the implementation. The protocol developed is observed and controlled, the mechanics of the application are adjusted, the design is carried out, the manuals and procedures are adjusted of the tests, the phases are improved and a first report with the previously registered information is delivered.

Therefore, this study is structured in a quantitative approach, because a set of systematic and empirical stages were applied to the study of the implementation plan of the application of this research [4]. The present work was developed through the structuring and execution of the following elements and investigative procedures:

- Preparation of spreadsheets for each technical variable.
- Pilot test to eliminate error factors.
- Internal validation of instruments.
- Collection of information.
- Quantification of data.
- Data systematization.
- Statistical treatment of data.
- Analysis of data and results of the information obtained.
- Approach of conclusions and recommendations.

The type of research is Evaluative due to the use of mechanisms that allow us to evaluate the level of technical effectiveness of the shot at goal of the soccer players under study. In addition, systematic and structured evaluation techniques of the events occurred were used. The analysis of these results allowed us to structure a characterization of the players under study and improve the sporting results in that direction, guidelines offered by the evaluative research [5].

2.3. Development
In this phase, the construction of the source codes is carried out, which is reflected together with the requirements and characteristics defined for each step, pilot tests are made, and controls or posttests are made that are necessary for the completeness of the system. A second report with the prototype of the application is delivered at the end of this phase and adjustments are made to the architecture of the test system.

From a classical perspective, technique is understood as the result of the cognitive and nervous processes of perception and tactical thinking, counting on the physical characteristics and skill level of the executor [6]. It is also interpreted as the mode of achievement of a motor objective, [7] indicated by tactics, taking into account the level of development of psychic and motor characteristics, highlighting the dependence on tactics and technique. In other words, it is the motor or visible phase of the game action.

It is understood then that these factors are interrelated, and it is easy to think that the skills of the athlete are revealed by physical qualities in sports. Therefore, it is understood the technique as a procedure, that leads directly and economically to achieve a high result. It is defined as an economic procedure or process; this term is being understood as the perfect realization between the motor project and its realization [8].

The sports technique is interpreted as the ideal model of competition action, developed on the basis of practical or theoretical experience in the athlete when performing the competition, which is characterized by such or such degree of effective use by athletes. This one, and of his possibilities to reach the sport result [9].

It can be proposed, then, that technical instruction would go through a first stage of a trend towards multilateral, multi-purpose development. A second one of general preparation of the chosen sports technique, and a third of special preparation or specialization, in which the chosen sport modality and the position or function that the athlete plays in his field of play or in the field of development of your sport.

The process of technical instruction could then be decomposed into four phases instead of three, as follows [10]:

1. Information and apprehension phase. The athlete takes knowledge of the movements to be learned and creates the necessary bases for the conception of an action project. At this point, his previous motor experiences, his motor level and his ability to observe help the subject.
2. Rustic coordination base. The first experiences of practical execution, such as verbal indications, represent the main information of this phase. At the end of this stage, the domain of the problem is in a rustic phase. The phenomena related to this level are: an excessive and partially erratic effort; the abruptness in the temporal development; the angular execution of movements; the insufficient amplitude; the false motor rhythm (very fast or very slow); the lack of rhythm and sequence and the lack of motor precision.
3. Fine coordination phase. The phenomena that characterize this stage are: the adequate energy cost, the necessary force expenditure, the amplitude, the rational motor rhythms, and the most fluid movements. The increase of the motor precision is here globally linked to a constant perfected understanding of the verbal or other information.
4. Consolidation, refinement and variable availability phase. In this phase, we find the successful coordination of movements, even in difficult or unusual conditions. Automation allows the athlete to focus attention on the critical points of motor development, and precisely the phenomenon that characterizes this stage is the constant and harmonious fluidity of the movements.

The achievement of these objectives allows developing more specific forms of training to avoid execution errors that start from the comprehension (cognition) that, the athlete has of its technical realization. In this line is the training based on the so-called Principle of clarification or discrimination by which tests are performed with artifacts of different weights, so that by kinesthetic differentiation, certain problems are solved, of execution, fundamentally in the application of forces.

It also gives the possibility of failure to recognize the real origin of errors in the execution of technical skills. In the same way, cognitive processes achieve the stability of the athlete with respect to the motor performance of the technique to achieve high performance in any circumstance, which is called "Constancy of movement" [11]. Cognitive processes also have a high participation in the elimination of the so-called "Speed Barrier" or "Speed Block" [12]. As well as other applications both in the competitive sphere, and as application in the training phases of technical skill.

Therefore, in the training process of the technique, both forms of motor participation should be used, as well as cognitive-affective participation, which complicates its application and lengthens it during practically the entire sports life of the subjects. It will be necessary to use different forms according to what the athlete needs to improve his technique, because each of these training options are more or less effective for the achievement of one or other technical objectives, which must be obtained in the course of the different stages of the formation of technical skill.

The teaching of football indicates that for a long time we have made our children play in imitation of adults. Moreover, for the young child, adult football is too "big", too difficult and complex because it requires a level of physical and mental abilities that children do not yet possess [13].

Instead of adapting the child to football, as it happens everywhere, we must know how to adapt football to the child, always to its perfect measure throughout its growth, like his shoes or his coat. Doctor Wein [13], an internationally recognized expert in team games who has dedicated himself for years exclusively to the important task of recycling the knowledge of football coaches, explains in a simple manner why and how of the need to adapt the game of soccer in all of its facets tailored to the child. In the first place, it is questioned whether children between 7 and 13 years old should compete in the regulation football, in imitation of the adults with eleven players in each team disputing a regulation ball in a field with giant dimensions and goals in which the goals are lost small goalkeepers.

I think he is right. If a child of seven or eight years still does not know how to dominate his body, how can he be expected to dominate the ball and one or more opponents? He is still exposed to exercise with a ball that reaches almost to the knee, a weight that his leg can't move easily and lack of time due to the presence of a large number of adversaries and colleagues around him, in search of This appreciated ball that everyone wants to have as long as possible.

Thus, a child comes to play three or four times the ball or a maximum of one minute in each game, completely insufficient to give him the necessary stimuli to be able to grow. Therefore, says Wein [13] in the aforementioned book that: "many players die with 33 -35 years on the football field, before they were born completely, by exposure to inappropriate methods, training contents and competitions during his best years of motor learning, between 8 and 12 years old".

2.4. Validation and evaluation

The In this stage we will show the results obtained in the pilot tests and we will talk a little about the application; this application is accessible in smartphones, tables and computers. Its architecture is designed in a JAVASCRIPT language by means of ANDROID studio and is connected to MYSQL database management systems. Also, how much with a security system of 3 types of user.

Super user: The first one is a user who can add clubs or teams and at the same time athletes to whom they want to evaluate, this same user will also have available to perform the evaluations.

Evaluating user: The second user is made for the evaluators, since it will only allow the player to be selected and he will be able to enter the data he takes during the evaluation.

The third user is made for people interested in knowing the data and it is designed to be accessed through the computer.

In this stage the expert judgments will be presented, and the previous stages will be verified in order to be able to carry out the final evaluation of the application.

3. Results

The results of this study are under construction. This work is a contribution to the scientific, academic and sports training community and will enrich the specialized literature.

This work will also allow to control and evaluate the technical gestures to be able to find faults in the movements of execution of the sports technique.

With the implementation of this work it will be possible to establish a modeling of the characteristics of the phases of sports techniques.

4. Conclusions

The model of evaluation of the level of technical effectiveness of the shot to goal developed in this application allows elite sports development. In addition, the correct development of this type of work offers equating and optimizing the technological and methodological tools that make possible the evaluation of the planning and organization processes of sports training.

References

[1] Seirul-lo Vargas F 2017 *El entrenamiento de los deportes de equipo* (España: Mastercede)
[2] Padrón G J 2006 *Bases del concepto de investigación aplicada* Consulte on: http://padron.entretemas.com.ve/InvAplicada/index.htm
[3] Arias E A, Jiménez J O, Olaya Z D and Restrepo L F 2003 *Valoración técnico-táctica de los deportes de cooperación oposición en los que participan niños de 10 a 12 años* (Medellín: Universidad de Antioquia)
[4] Hernández F y Fernández C 2006 *Metodología de la investigación* (México: McGraw Hill)
[5] Tamayo M 1999 *Serie aprender a investigar. módulo 2 y 5, La investigación y el proyecto de investigación* (Bogotá: ICFES)

[6] Mahlo F 1974 *La acción táctica en el juego* (La Habana: Editorial Pueblo y Educación)
[7] Theodorescu L 1976 *Teoría y metodología del deporte* (Paris: Editorial Francaise Reunis)
[8] Mata F 1993 *Consejos en la planificación del trabajo del fútbol Base* (Barcelona: Editorial Hispano-Europea)
[9] Matveev L 1977 *Periodización del entrenamiento deportivo* (Moscú: Editorial Raduga)
[10] Weineek J 1995 *Manual de entrenamiento deportivo* (Barcelona: Editorial Paidotribo)
[11] Meinel K, Schnabel G and Krug J 1988 *Teoría del movimiento, motricidad deportiva* (Argentina: Editorial Stadium)
[12] Osolin N G 1983 *Sistema contemporáneo del entrenamiento deportivo* (Ecuador: Editorial Científico Técnico)
[13] Wein H 1996 *Fútbol Base Programas de Entrenamiento 14-15 años* (Barcelona: Editorial. Paidotribo)

IOP Conf. Series: Journal of Physics: Conf. Series **1161** (2019) 012018 doi:10.1088/1742-6596/1161/1/012018

Impact of cultural transformation campaigns around energy savings in hotels. Study in Bucaramanga and Valledupar, Colombia

S Serrano[1], Y Muñoz[2] and J Quiroga[3]
[1] Facultad de Psicología, Universidad Antonio Nariño, Bucaramanga, Colombia
[2] Departamento de Electromecánica, Universidad Antonio Nariño, Bucaramanga, Colombia
[3] Escuela de Ingeniería Mecánica, Universidad Industrial de Santander, Bucaramanga, Colombia

Email: sandserrano@uan.edu.co, yecidmaldonado@uan.edu.co

Abstract. The study focused on evaluating the impact of an educational campaign aimed at transforming attitudes and practices around the saving of water and energy in regular customers of two hotels, located in two touristic cities of Colombia, with very different idiosyncrasy: Valledupar and Bucaramanga. The development of the campaign was framed in the principles of the rational and efficient use of Energy Program (PROURE), and addressed three areas: information, the promotion of rational use practices and, increase in the awareness of the care of the environment. The proposed design was quasi-experimental, pretest / posttest. The sample was for convenience, involving a total of 180 people among employees and customers (100 in Bucaramanga and 80 in Valledupar). Participants responded to a survey of attitudes and practices, before and after participating in the campaign that lasted for 6 months. The results indicate that in the city of Bucaramanga the changes obtained in the posttest were significant, around the reflection and awareness (attitudes) of the saving of water and energy, by the guests. In the case of Valledupar, there are more changes in the employees than in the guests, especially in the attitudes they have regarding the importance of energy saving. In general, the study made it possible to demonstrate the importance of programs and processes of education and culture that promote these attitudes and practices which can result in significant impacts to the care of the environment.

1. Introduction

The concern for issues related to the care of the environment and the promotion of a culture of care is recent in the world, in general and particularly, in Colombia [1]. Until a few years ago, policies, the design of educational campaigns and the promotion of saving behaviors have begun to be thought in function of a necessary cultural transformation, because in the country, the subject seemed to be far, given the abundance of natural resources and little social awareness to practices such as recycling, saving water or controlling energy consumption [2,3]. It can be said that for years, people have been related in an unaware and even irresponsible way to the natural resources.

Consequently, families, communities, cities and businesses are not focused on the promotion of water and energy saving habits, or on the dynamics of consumption [4]; perhaps because there have been other priorities and needs of poverty and limited socio-economic conditions.

IOP Conf. Series: Journal of Physics: Conf. Series **1161** (2019) 012018 doi:10.1088/1742-6596/1161/1/012018

This article presents the results of a joint investigation on the promotion of a culture of water and energy saving, in two hotels, in the cities of Bucaramanga and Valledupar, Colombia.

The existing literature about water and energy saving is related to three of the most important characteristic aspects of rational and efficient consumption: Efficient use of energy as an alternative to improve competitiveness and productivity in companies, promotion of sustainable consumption practices of resources and articulation of campaigns to promote sustainability.

Amell [5], Arroyave & Garcés [6] and Deming [7] addressed the efficient and rational use of energy as an alternative to improve competitiveness and productivity in companies. In Colombia Tobasura, Moreno, Aya & Mora [8] conclude in a study of energy and financial productivity in coffee farms, that there is a direct relationship between these two variables, which contradicts the belief that industrialized and more specialized production systems are more financially efficient. Hartley [9] in Argentina defines that Eco-efficiency as an administrative philosophy that drives productive activities in the search for environmental improvements and economic benefits [10]. Efficiency, in this sense, is approached from the use of natural resources, under the premise of producing more with less.

Salgado-Beltrán and Beltrán-Morales [11] investigated the factors that influence the sustainable consumption of organic products in Mexico; consumer behavior (health behaviors), and demographic aspects (schooling) exert a positive effect on sustainable consumption, which shows a directly proportional relationship between these two variables: marketing strategies (ecological communication and price) encourage the sustainable consumption.

In relation to the development of communicative campaigns to promote sustainable consumption, the United Nations Environment Program [12,13] highlights the importance of these campaigns in the promotion of sustainable development and the need to make them accessible and understandable. For the success of the same, the content, communicator, tone and means of communication are fundamental; highlighting the determining factors for the effectiveness of the campaigns and the need for studies that can define the associated factors [14,15].

In the Latin American context, the Energy Efficiency Network has been created, which brings together the agencies of all countries, facilitating the exchange and dissemination of technical, legal and regulatory information among interested institutions and professionals. Some of the member countries have made important advances in the development of programs for Rational Use of Energy (RUE) implemented methodologies and promotion campaigns [16,17].

In Colombia there is not a policy that integrates the energy efficiency programs, the actions developed have been conjunctural, without been coordinated by a central entity that channels them[18]. Sustainable production and consumption policies cover aspects such as urban environmental management, national biodiversity policy for the development of ecotourism, environmental education, among others, but it does not refer to an energy efficiency policy [19].

In Colombia the development of RUE has been relatively inefficient, being necessary to determine the weaknesses of policies and programs to propose new strategies [20,21].

2. RUE approximations

By efficient use of energy is understood the improvement in the capacity of the processes of utilization and transformation without needing to reduce the demand of useful energy. Energy saving means a reduction of the final energy demands caused by net reductions in the useful energy requirements, that is, of the energy service requirements of the society and the industry. The term rational use of energy sometimes covers the above aspects plus the creation of energy sources with a low percentage of carbon [22].

2.1. The attitudes and the attitude change

Attitudes are our evaluating feelings, are affective or emotional and refer the favorable of our feelings towards a particular objective. In that sense, an attitude is a state of mind. If we want to influence the behavior of others, one way to do this is to influence the mood of the people [23].

Social Psychology defines three components of attitude: the cognitive, affective and behavioral element [23,24]. Attitudes are powerful engines that provide the energy for the behavior and direct it. But this is not always the case; many studies reveal a lack or low correspondence between verbally expressed attitudes and behavior [25].

The attitudes that we form on the basis of our direct experience are more reliable for predicting our actions than those that we form indirectly [26] maintain that although the attitude we have towards an object influences our general pattern of response to it, it does not allow us to predict our specific action with respect to the object [27].

Attitudes towards the energy problem are not so related to informed conservation behavior but specific beliefs and personal consequences of using less energy [28].

The communicative campaigns for attitudinal change and, in short, behavioral, consider not only what is going to be communicated, but also how to communicate it. Among the important factors for success are the content, the communicator, the choice of tone and the media [12].

3. Methodology

A quantitative study with a descriptive scope was carried out. The research design used was quasi-experimental, with no control group, pretest / posttest [29]. A convenience sample was used, consisting of 180 employees and clients of two hotels, one located in the city of Valledupar and the other in the city of Bucaramanga; two touristic cities of Colombia, with very different idiosyncrasy, (the sample was 100 in Bucaramanga and 80 in Valledupar).

The instruments that were used were two: a survey of attitudes and practices, instrument designed for the researchers, and validated by two experts in instrument validation , which contained aspects of information, care practices and attitudes in relation to savings culture and, a campaign to promote attitudes and practices around saving energy and water in the hotels [30]. This campaign included audiovisual material, informative spaces that were implemented during 6 months in each hotel.

4. Results

The data about the employees of the hotel in Bucaramanga, showed that there is no significant difference in the variables attitudes and practices pre and post-test; based on the results of the application of the *t-student* statistic (suitable for the size and distribution of this sample) with the software EZanalyze v. 3.0, and applying the concept of p-value or probability value or asymptotic significance, with a reference value of significance of 0.05; with p values of 0.64 and 0.92 respectively, higher than the reference value. These results are shown in Tables 1 and 2.

Table 1. Results report - paired t-test of pre-e/b attitudes with post-e/b attitudes.

	Pre-E/B ATtitudes	Post-E/B ATtitudes
Mean:	83.889	82.111
Std. Dev.:	12.683	5.134
N Pairs:	9	
Mean Difference:	1.778	
Se Of Diff.:	3.730	
Eta Squared:	0.025	
T-Score:	0.477	
P:	0.646	

The analysis of results in the Bucaramanga guests, starting shows a significant difference in the Attitudes, with a value of p of 0.03, lower than the reference value of 0.05; which reveal an increase of the variable in the post test, meaning a positive impact of the intervention in this sample of the study population, Figure 1 illustrates the results found and Table 3 the results of the applied statistic.

Table 2. Results report - paired t-test of pre- e/b practices with post- e/b practices.

	Pre- E/B PrActicEs	Post- E/B PrActicEs
Mean:	84.125	83.750
Std. Dev.:	9.877	11.474
N Pairs:	8	
Mean Difference:	0.375	
Se Of Diff.:	3.826	
Eta Squared:	0.001	
T-Score:	0.098	
P:	0.925	

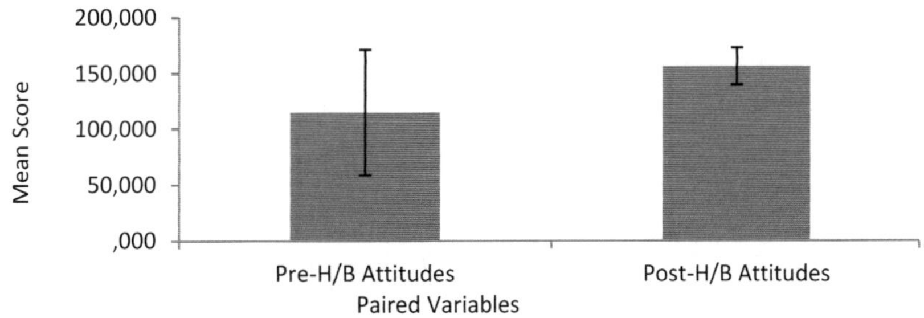

Figure 1. Comparative Pre-test/Post-test Attitudes for guests in Bucaramanga.

Table 3. Results report - paired t-test of pre-h/b attitudes with post-h/b attitudes.

	Pre-H/B Attitudes	Post-H/B Attitudes
Mean:	114.889	156.222
Std. Dev.:	56.412	16.724
N Pairs:	9	
Mean Difference:	-41.333	
Se Of Diff.:	15.739	
Eta Squared:	0.434	
T-Score:	2.626	
P:	0.030	

The study at the hotel in Valledupar reveals the following results. In the analysis of the employees of the hotel against the variable Attitudes, a significant difference is shown with a tendency to a decrease of the values for the arithmetic averages in the post-test, whose pre-test means were 162.1 and in the 147.6 post-test, revealing an inverse impact of the intervention performed; these data are illustrated in Figure 2 and Table 4.

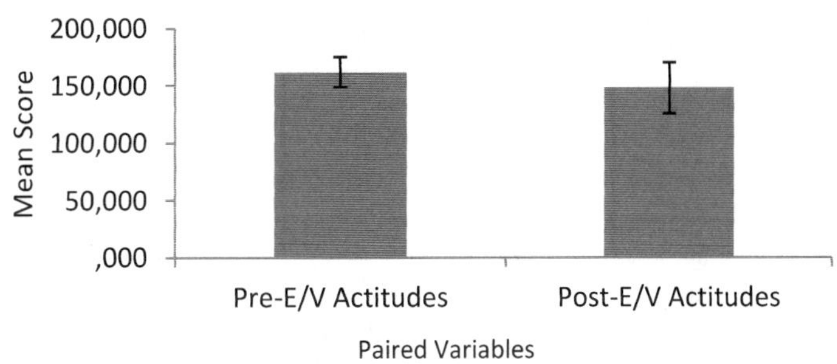

Figure 2. Comparative Pre-test/Post-test Attitudes for the employees in Valledupar.

In these results, the means found were, in pre-test 162.1 and in post-test 147.6, the value of the variable Attitudes in the post test being lower; The statistical results are shown in Table 4.

Table 4. Results report - paired t-test of pre-e/v attitudes with post-e/v attitudes.

	Pre-E/V ATtitudes	Post-E/V ATtitudes
Mean:	162.111	147.667
Std. Dev.:	13.280	22.389
N Pairs:	9	
Mean Difference:	14.444	
Se Of Diff.:	5.556	
Eta Squared:	0.429	
T-Score:	2.600	
P:	0.032	

Regarding the variable Practices for the hotel employees of the hotel in Valledupar and the guests variables, the results do not reveal a statistically significant difference. Table 5 shows the results for the guests sample in Valledupar.

Table 5. Descriptive statistics, Valledupar guests.

	ATtitudes		PrActicEs	
	Pre-test	Post test	Pre-test	Post test
Mean	333.9	337.0	322.0	336.3
Std Dev.	17.5	20.2	30.3	34.1

5. Conclusions

The results of the study indicate that the campaign to promote a culture of saving energy and water in hotels, achieved significant changes in the attitudes that the guests of the Bucaramanga hotel showed. However, the practices did not show statistically significant changes after the campaign was applied. This is explained from the general theory of attitudes, since from Social Psychology the authors have shown how to obtain a general positive attitude towards a specific issue or situation (in this case the culture of saving water and energy), not necessarily accompanies practices that are coherent with these attitudes.

About the campaign in the hotel of Valledupar, the results confirm that attitudes can be transformed, but it is necessary to continue promoting concrete actions around an issue such as the culture of savings as the practices of both, employees and guests, did not show significant changes after participating in the campaign.

It is important to note that the applied pedagogical campaign addressed aspects related to attitudes and practices in relation to a water and energy saving culture, in an area such as hotels; although statistically the changes presented were not significant in the four groups (employees of the two hotels and guests of the two hotels), there were changes between the pre-test and post-test, although it cannot be stated categorically that the intervention was effective, it generated changes, meaning that attitudes and practices changed positively between the pre-test and the post-test.

References

[1] Vanegas J and Botero S 2012 Eficiencia energetica en microempresas de medellin: un estudio de valoracion de barreras *Lecturas de Economia* **77** 1-33

[2] Gaona Pérez A 2000 Desarrollo sostenible y desarrollo solidario *Comununicar* **15** 83–91

[3] Alonso L E, Fernández Rodríguez C J and Ibañez Rojo R 2011 Del consumismo a la culpabilidad: en torno a los efectos disciplinarios de la crisis económica *Política y Sociedad* **48** 353–379

[4] Perez J A, Rosenkranz F, Chamy L and Elortegui Raffo2014 *Proyectos de eficiencia energetica para institucuones de educacion superior* (Santiago de Chile: AChEE)

[5] Amell A 2007 *Energía, economía y medio ambiente en procesos industriales* (Bogota: Universidad Nacional de Colombia)

[6] Arroyave Rojas J A and Garcés Giraldo L F 2006 Tecnologías ambientalmente sostenibles *Prod. + Limpia* **1(2)** 78-86
[7] Deming W E 1989 *Calidad, productividad y competitividad: la salida de la crisis* (Madrid: Ediciones Díaz de Santos S. A.)
[8] Tobasura I, Moreno F A, Aya S and Mora J 2012 Productividad energética y financiera en fincas campesinas del departamento de Caldas. Tres estudios de caso. *Revista Luna Azul* **34** 101–12
[9] Hartley Ballestero R 2009 Eco-eficiencia una gestión empresarial ambiental *Revista Ciencias Económicas* **27(1)** 189-205
[10] Secretaria de Planeamiento Energético 2014 *Campaña Hagamos Click* (Argentina: Ministerio de Energía y Minería)
[11] Selgado Beltran L and Beltrán Morales L 2011 actores que influyen en el consumo sustentable de productos orgánicos en el noroeste de México *Universidad y Ciencia* **27** 265–79
[12] Futerra 2005 *Comunicando la sustentabilidad. Cómo producir campañas públicas efectivas* (Kenia. Programa de las Nacionas Unidas para el Ambiente)
[13] Prias Caicedo O F 2010 *Programa de uso racional y eficiente de energía y fuentes no convencionales – PROURE* (Bogota: Ministerio de Minas y Energia)
[14] Rubio M C 2010 UREE: Campaña de Uso Racional de la Energía Eléctrica implementada en la provincia de Río Negro *Voces en el Fenix* **2(10)** 44-49
[15] Aponte Gutierrez J C and Andrade Mahecha J F 2013 *Proyección de Demanda de Energía Eléctrica en Colombia* (Colombia: Ministerio de Minas y Energía)
[16] CORPOELEC 2013 Uso Racional y Eficiente de la Energía Eléctrica. Importancia de las políticas de uso racional y eficiente de la energía. Consulted on: http://www.corpoelec.gob.ve/uso-racional-y-eficiente-de-la-energia-electrica
[17] Peréz S 2013 Propuesta Metodológica de acción participativa para promover la intervención de la comunidad en la solución de los problemas ambientales. Caso de estudio: Trujillo - Venezuela *Agora Trujillo* **15(29)** 135-151
[18] Convenio UPME-MEN 2009 *Documento borrador. Introducción de la temática de energía y su uso racional en el programa educativo medio ambiental en Colombia* (Colombia: congreso de la Republica)
[19] Dirección de Desarrollo Sectorial Sostenible 2011 *Política Nacional de Produccìon y Consumo. Hacia una cultura de consumo sostenible y transformación productiva* (Bogota: Ministerio de Ambiente, Vivienda y Desarrrollo Territorial)
[20] Consorcio Bariloche – BRP 2007 *Consultoría para la formulación estratégica del plan de uso racional de energía y de fuentes no convencionales de energía 2007 – 2025* (Bogotá: Fundacion Bariloche)
[21] Bessudo S 2011 Acciones para la Sostenibilidad Ambienta. El gobierno articula iniciativas de Eficiencia Energética. *Mecanismos e Instrumentos Financieros para Proyectos de Eficiencia Energética en Colombia* (Bogota: Ministerio de Minas y Energía) pp 12–3
[22] COLCIENCIAS 2005 *Plan estratégico programa nacional de investigaciones en energía y minería. Bases para una política de promoción de la innovación y el desarrollo tecnológico en Colombia* (Bogota: Colciencias)
[23] Vander Zanden J W 1986 *Manual de psicología social* (Barcelona: Paidós)
[24] Perlman D, Cozby P C and Soto Suárez J N 1969 *Psicología social* (México: McGraw-Hill)
[25] Wicker A W 1969 Attitudes versus actions: The relationship of verbal and overt behavioral responses to attitude objects *Journal of Social* **25(4)** 41-78
[26] Abric J C, Moscovici S and Ibáñez Gracia T 1985 *Psicología social* (Barcelona: Paidós)
[27] LaPiere R T 2010 Attitudes vs Actions *International Journal of Epidemiology* **39** 7–11
[28] Rossetti R and Giacomini Filho G 2010 Comunicação, consenso social e consumo sustentável *Comunicacao Midia e Cosumo* **7** 153–69
[29] Hernández Sampieri R, Fernández Collado C and Baptista Lucio P 2014 *Metodología de la investigación* (Mexico: McGraw-Hill)
[30] Navarrete Cobo B 2011 La gestión sostenible y su comunicación en el sector hotelero de la ciudad de Málaga *Prisma Social* **6** 30-62

IOP Conf. Series: Journal of Physics: Conf. Series **1161** (2019) 012019 doi:10.1088/1742-6596/1161/1/012019

Online training for the strengthening of mathematical pre-knowledge mediated by Khan-Academy platform

K Rueda[1] and O Serrano[1]
[1] Grupo de Investigación en Ciencias Básicas y Aplicadas, Unidades Tecnológicas de Santander, Bucaramanga, Colombia

E-mail: krueda@correo.uts.edu.co, sjaimez@correo.uts.edu.co

Abstract. The absence of mathematical skills makes part of the formative challenges currently facing Colombian Higher Education Institutions (HEIs) Being consequently Aware of the situation, denoting from the second academic semester of 2018 the Technological Units of Santander have consistently implemented the free course PREIN-UTS. This is an online training course mediated by the artificial learning web platform Khan-Academy, and it's directed at the first semester students with the aim of strengthening their mathematical pre-knowledge before they are accessing the institution. Therefore, the object of study is to analyze the implementation of the course in his first stage and how the students does use it to improve their learning in mathematics. The methodological approach is quantitative of exploratory type with the proper use of descriptive statistics in the SPSS software. The random sample was made up of new first semester students (N=277). The results show implicitly the manner of solving exercises to increase the level of performance of the skills, in fact confirms and reveals that the activity students that dedicates the most time with average 340.97minutes and Standard deviation (SD) 413.27. Finally, it was concluded that Khan-Academy enables the creation of free easy-access courses that can be used for the benefit of students who want to get into Higher Education to encourage their mathematical skills.

1. Introduction

Information and communication technologies have dabbled on a large scale in the educational field with the generation of several tools that will brought dynamism in front of learning and knowledge [1,2] by promoting the formative processes of students at all educational levels [3,4]. In the other hand, when the Colombian applicants try to get access to the higher Education they eventually present great shortcomings and difficulties in the development of mathematical competences [5,6], which generates high dropout and repetition rates around the first academic semesters [7], emphasizing within the current challenges that must be addressed by the different Colombian Institutions of Higher Education (HEIs) [8]. Precisely, the Technological Units of Santander ever since the initiative was consider in the second half of 2018 have launched the online training course called PREIN-UTS that pretends the strengthening of mathematical skills mediated by the Khan-Academy technology platform.

Khan Academy is a digital tool that offers different free courses in which the tutors have direct access to detailed information of the work carried out by their students in order to identify difficulties in time and implement improvement strategies to overcome the obstacles of learning [9]. The Khan-Academy training courses contain various methods and different support of aids that enable self-learning through videos, missions, clues and readings [10,11]. Besides, the platform rewards the achievements obtained through medals, avatar and energy points, which motivates the student to fulfill their goal [12]. In the same way, Khan-Academy continuously assesses the level of skill development on a qualitative scale as follows: With difficulties, needs more practice, practiced, level 1, level 2 and Mastered.

Therefore, this study describes the experience of the implementation of the first phase of the PREIN virtual course mediated by the web platform Khan Academy conscientiously in students who were in fact admitted pursuing the first semester in to the different academic programs of (HEIs) of Bucaramanga, Colombia.

2. Methodology

The methodological approach is quantitative of exploratory kind with the use of descriptive statistics in the SPSS software. The random sample was shaped by new students of first semester (N=277). The research was actually separated in the following study phases:

2.1. First phase

A video tutorial was designed, and it was originally created the PREIN course by the Khan-Academy corresponding with the respective internal codices. The thematic axes assigned from the Khan-Academy were the primal bases of algebra according to: Unity 1 Fundamentals (31 skills), Unity 2 Algebraic expressions (10 skills), Unity 3 Linear equations and inequalities (14 skills) Unity 4 line graphs, and slope (13 skills), Unity 5 Systems of equations (6 skills), Unity 6 Expressions with exponents (11 skills) Unity 7 Quadratics and polynomials (12 skills), Unity 8 Geometry (6 skills). For a weighted total of 103 skills.

2.2. Second phase

*F*rom the stringent institutional documental base some reports were generated properly taken from the explicit listing of the students of first semester. Subsequently, the intention of create an invitation purpose was emitted via email and telephonically

2.3. Third phase

*T*he preliminaries results were analyzed consider the following variables of study:

- Descriptive variables: gender, headquarters, academic program and faculty.
- Khan-academy Variables: skills, makes reference to the exercise practice inside the platform environment; activities, it does refer to the spender time within the platform; and gamification, these are the key points of energy and medals earned.

3. Partial Results

3.1. Descriptive results

Of the 1624 students enrolled in the first academic semester 277 were registered on the PREIN course. Whom which the 53.36% were man ant the el 46.64% were woman. The Figure 1 show the percentage distribution of PREIN students by headquarters where it gets a great appreciation of a major participation of the principal headquarters students of the institution with a 91.18%.

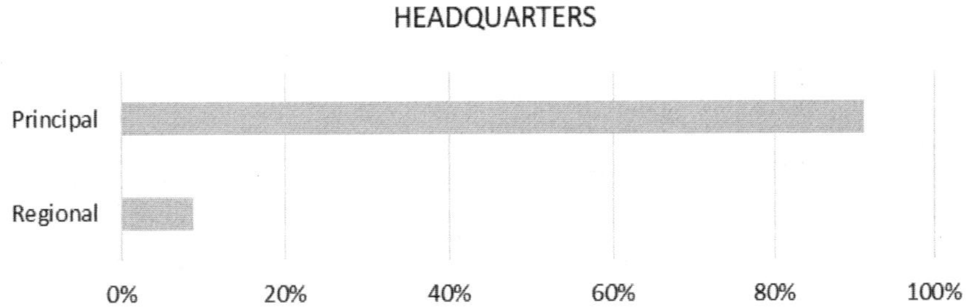

Figure 1. Bar graph of PREIN students by headquarters.

By a side, the Figure 2 sights the distribution of the percentage of students per academic program: Tourism, Geotechnics, Agroindustry engineering, Topography, Fashion, Electronics, Telecommunications, Banking and finance, Electricity, Sports, Environmental engineering, Marketing, Systems, Electromechanical engineering, Business management, accounting career; highlighting the participation of students coming all the programs offered by the institution.

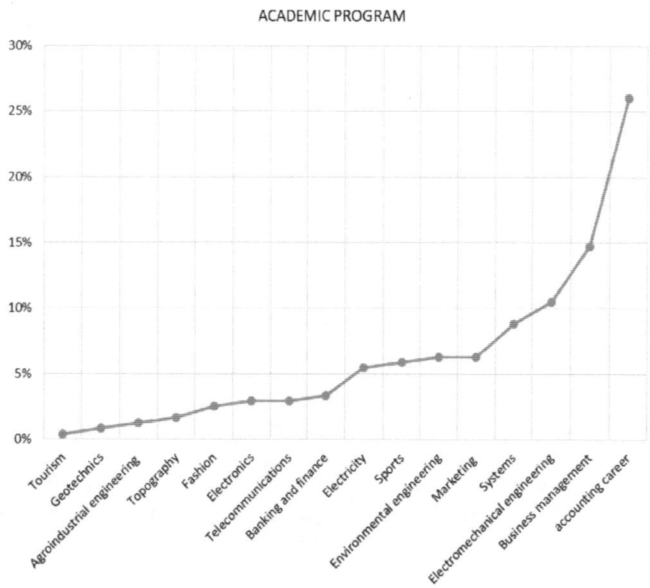

Figure 2. Line graphic of PREIN students by academic program.

In the other hand, the sectoral graphic from the Figure 3 evidence a considerable increase of participation from the belonging students that attend at the Faculty of socioeconomic and business sciences in reference to the students of the Faculty of Natural Sciences and Engineering.

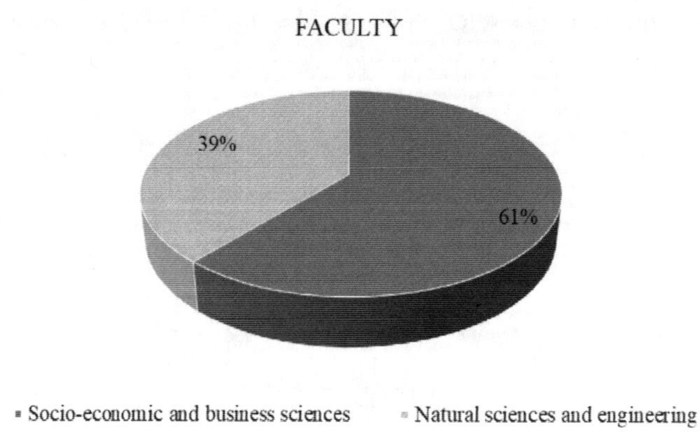

Figure 3. Sectoral graphic of PREIN students per faculty.

3.2. Khan-Academy results

3.2.1. Abilities. After a month of the implementation in use of the PREIN course its absorbs that the 40% of the student to have a reliable progress major or equal to the 50 % in the exercise practice (See Figure 4).

3.2.2. Activity. Respect to the invested time it is planed that the students have, on average, spent 444.08min on the platform, distributed as follows: 103.11min watching video tutorials and 340.97min to practice missions (See Table 1) From the above we can conclude and affirm certainly that students prefer to invest notoriously their time in solving exercises than observing videos.

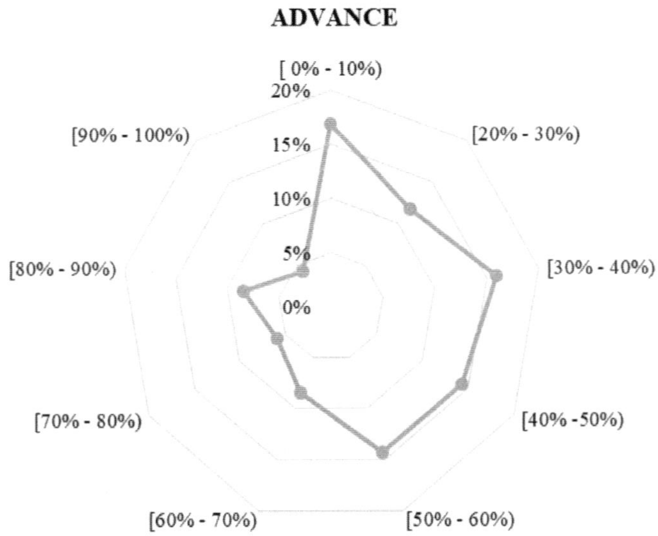

Figure 4. Radial Graphic of the global advance of the students in PREIN.

Table 1. Descriptive statistics.

Total minutes		Minutes of videos		Minutes of skills	
Mean	444.08	Mean	103.11	Mean	340.97
Median	279.63	Median	38.57	Median	229.07
Mode	#N/A	Mode	0.00	Mode	#N/A
Standard deviation	553.41	Standard deviation	199.41	Standard deviation	413.27
Sample variance	306262.08	Sample variance	39762.85	Sample variance	170789.68
Kurtosis	26.80	Kurtosis	25.82	Kurtosis	32.91
Asymmetry coefficient	4.49	Asymmetry coefficient	4.60	Asymmetry coefficient	4.88
Minimum	60.05	Minimum	0.00	Minimum	50.13
Maximum	4306.92	Maximum	1461.67	Maximum	3416.28
Confidence level (95.0%)	113.35	Confidence Level (95.0%)	40.84	Confidence Level (95.0%)	84.65

Right Next, Figure 5 demonstrate the Box Plot of the total amount of minutes students have invested in the platform, where we absolutely can objectively appreciate the asymmetry inside the distribution of the variable, maximum value of 4306.92min, minimum value of 60.05min and median of 279.63min.

3.2.3. Gamification. In fact, so far at this moment the average energy points obtained by students on the Khan-Academy platform is 44250 with a Standard Deviation (SD) 39671.53 while the number of medals won is 17 (SD) 9.40.

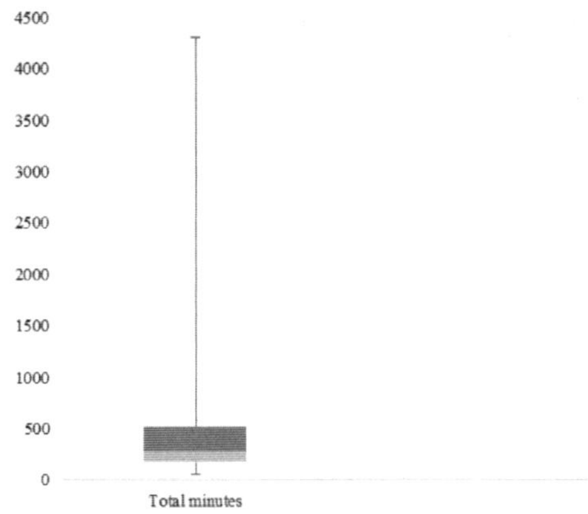

Figure 5. Box Plot of the total time measured in minutes on the KHAN-Academy platform.

4. Conclusions

The accurate technology has entirely made it possible for the different location students, their permeable access to the PREIN course presetting without the prevailing need of moving to the main Headquarters and actually incur in different mobility expenses, allowing access to the platform from the comfort of their homes or making use of any site that counts with a viable computer and a good internet connection. The matter fact being implicit that Khan-Academy is a free software does incurred prioritizing in the minimum expenses generated for the institution and clearly does it make that the current implementation of the PREIN course be enormously significant, contributing to the valuable creation of teaching innovative strategies that promote the mathematical pre-knowledge of the students that will get in to first semester of any academic program. Even though the PREIN is a voluntary course, in this first stage of implementation it had a great acceptance by the students of the two faculties. Finally, in the next implementations is recommended extend the invitation to the students repeaters of the first semester subjects related to the area of mathematics.

References

[1] Roslina C, Zarli M, Mawengkang H, and Sembiring R 2017 The role of information and communication technology in developing smart education *Journal of Physics: Conference Series* **890** 012091

[2] Batmetan J and Palilingan V 2018 Higher education students' behaviour to adopt mobile learning *IOP Conference Series: Materials Science and Engineering* **306** 012067

[3] Utomo H, Bon A and Hendayun M 2018 The integrated academic information system support for education 3.0 in higher education institution: Lecturer perspective *Journal of Physics: Conference Series* **1049** 012102

[4] Castrillón M, Murillo E, Moreno J and Polo J 2017 Estudiantes de la básica y media con respecto al uso de las TIC como herramientas de apoyo a su aprendizaje *Trilogía Ciencia Tecnología Sociedad* **9** 113

[5] Murcia M and Henao J 2015 Educación matemática en Colombia, una perspectiva evolucionaria *Entre Ciencia e Ingeniería* **9** 23

[6] Torres C, Peña T, Sola M, and Lopez N 2017 Análisis de los factores de la deserción estudiantil en una facultad de las Unidades Tecnológicas de Santander, Bucaramanga, Colombia *Notandum* **30** 141

[7] Barragán D and Patiño L 2016 Elementos para la comprensión del fenómeno de la deserción universitaria en Colombia. Más allá de las mediciones *Cuadernos Latinoamericanos de Administración* **9** 55

[8] Montes F, Forero D, Ricardo S, and Zarama R 2017 The Challenges of creating a ranking: a colombian example *International Higher Education* **90** 22

[9] Daza C and Simanca F 2016 Enseñanza de la derivada mediada por objetos de aprendizaje *Revista Vínculos* **13** 159

[10] Espinosa J, Marín L, Alzate H and Osorno N 2017 La incidencia de los objetos de aprendizaje interactivos en el aprendizaje de las matemáticas básicas, en Colombia T*rilogía Ciencia Tecnología Sociedad* **9** 63.

[11] Radcliffe D, Knappenberge T and Daigh A 2016 Using Khan Academy videos in flipped classroom mode to bolster calculus skills in soil physics courses *Natural Sciences Education* **45** 1

[12] Patrick K and Teomara R 2017 Khan Academy as supplemental instruction: A controlled study of a computer-based mathematics intervention *The International Review of Research in Open and Distributed Learning* **18** 70

IOP Conf. Series: Journal of Physics: Conf. Series **1161** (2019) 012020 doi:10.1088/1742-6596/1161/1/012020

Google classroom as a tool-mediated for learning

K L Cristiano[1] **and D A Triana**[1]
[1] CIMBIOS, Universidad Industrial de Santander, Bucaramanga, Colombia

E-mail: karen.cristiano@saber.uis.edu.co, dantrica@saber.uis.edu.co

Abstract. Abstract. One of the biggest matters which has a teacher at the University is how to find the most efficient way to administer their courses? With the aim to the students will interest and motivated to learn. The web service google classroom was included physics III laboratory and mechanics II laboratory at the Universidad Industrial de Santander as an alternative to the Moodle platform, to give students another point of view about learning in the physics laboratories and those topics traditionally studied through blackboard. Google classroom has been accompanied by a methodology of continuous monitoring, stimuli, activities based on problem-solving and assignments to improve their writing skills. Also, google classroom created courses were easy to follow. The methodology proposed could be extrapolated to other experimental or theoretical subjects. In addition, google classroom showed to be friendly to the environment.

1. Introduction

The dropout rate of university students exceeds 50% in Colombia, the largest dropout populations are in the first semesters [1–3]. Then, the Colombian Ministry of Education promotes the use of Information and Communication Technologies ICTs [4,5] as a pedagogical and didactic tool to connect students with those crucial sensitive experiences for their learning [6]. Such that, the old form of blackboard teaching and the books could not teach [7]. Due to the exposed above, the new teaching-learning strategies use ICTs in the study of physics [8,9]. Another initiative to be considered as an example to follow from the University of Colorado, they develop interactive simulations so that students can learn by exploring. The key is which the online platform to be intuitive enough in order to the student can advance their learning independently [10]. The culture of the modern student is based on smartphones use [11]. They like writing and reading content on social networks. It's seemed a problem, we have turned it into an opportunity to mediate the teaching-learning process. We have selected the web service google classroom to implement the strategy Just in Time Teaching (JiTT) [12] and other elements of mediated learning to improve transversal skills, such as reading and writing [13]. Through the google classroom, we haven't only adapted to the modern student, but we have managed to involve students with the contents of their interest and writing of short texts such as essays, abstracts, and reports.

We observed which the students were satisfied with the teaching accompaniment through the web service google classroom. They commented google classroom is nice to the user, the flow of activities is easy to follow, and activities are relevant and captivate the students. The concept of grade goes into the background prioritizing the expectation to learn.

2. Methods

Nowadays, at the Universidad industrial de Santander has institutionalized the strategy (JiTT) with elements of mediated learning through the learning platform Moodle, in the physics I, II and III laboratories [8].

The physics laboratory courses for physics students offered in the academic calendar and laboratory courses for engineer students offered on vacations don't have coverage of the learning platform Moodle for the implementation of the (JiTT). Therefore, in this work we implemented (JiTT) and mediated learning elements with the google classroom web service. The selected subjects to this study were physics III laboratory oriented to engineering students and mechanics II laboratory oriented to physics students.

The grade from physics III laboratory has a value of 20% within the theoretical subject. Both are considered as a single subject for engineering area. On the other hand, the laboratory subjects for physics students are independent.

2.1. Methodology for engineering students

The web service google classroom is divided into three tabs: stream, people and information. The tabs were used as follows. The students were invited through their email accounts to join the class. Their contact information was stored in the people tab.

In the information tab, the initial guidelines of the course were uploaded, such as programming of the laboratory sessions, activation dates of the preparation laboratory questionnaires, reports delivery dates, tutorial videos for use CASSY Lab software, instructions before laboratory sessions, worksheets and finally the research projects.

Laboratory preparation questionnaires were published in the stream tab and it's were enabled 28 hours before the laboratory session. The students did have 5 days to present their report. The assignments were uploaded for the students to submit each one of the reports. Finally, a research proposal on one of the studying issues in the laboratory to be proposed of recyclable materials and low cost was published.

The questionnaires were developed on google forms. Each questionnaire had four questions, two about the conceptual framework, one about the experimental methodology and one of the expected results. The percentages were, questionnaires 30%, reports 40% and research proposal at end course 30%.

2.2. Methodology for physics students

The mechanics II laboratory three hours per week for seventeen weeks was taught. The other hand, the laboratory for engineers in four weeks was taught, daily two hours. The physics students had more time to develop the activities.

Before the laboratory session in addition to the questionnaires, each student had to write an abstract content as follows: keywords, experimental methodology and expected results. They should write an essay for the technical visit to the steel plant, where metallographic preparation, tensile-compression tests, and hardness-microhardness tests were made. The students had fifteen days a deadline to submit the essay. Also, the reports had two weeks for deadline. The evaluation percentages are detailed in Table 1.

Table 1. Mechanical II laboratory grading.

Assignments	Percentage over grade
Laboratory preparation questionnaires	15%
Abstracs	15%
Essays	10%
Reports	40%
Final investigation proposition	20%

2.3. Measurement of the methodology

The Likert items with eleven questions was developed to evaluate the opinion of the students regarding the use of google classroom in laboratory subjects. The questions in three groups were separated: one, about used of google classroom (U), course resources (R) and alumni-teacher communication (C). The statements and questions formulated are shown in the Table 2.

Table 2. Likert items.

Group	Items
Have I used learning platforms other than Google classroom in physics? I think that google classroom is attractive, pleasant and easy to follow. The instructions on google classroom was clear. I would like to use google classroom to another practice and theory subjects. The objectives presented by the teacher was achieved.	U
The assignments posted on google classroom allowed me to improve the way learning I'm. I needed more explanation in order to understand the assignments. The course materials were according to the contents of the subject.	R
The assignments and feedback published thought google classroom were clear and coherent. I think to submit assignments and reports by mean google classroom was eco-friendly. Have I used to google classroom to make questions, class comments or announcements?	C

3. Results

The engineering and physics laboratory courses were taught using mediated learning through the google classroom web service. Every week, the teacher submitted assignments, questionnaires, comments, questions, assignments feedback; and resources such as scientific papers and documents to develop the experimental configuration named research projects, worksheets, videos and software. The other side, students submitted reports, essays, abstracts, questionnaires and interacted with the teacher leaving their questions as comments. At the end of the course, the students filled Likert items (see Table 2).

The opinion of engineering and physics students regarding the implementation of the methodology proposed here was shown in Figure 1. The results were grouped as seen in Table 2.

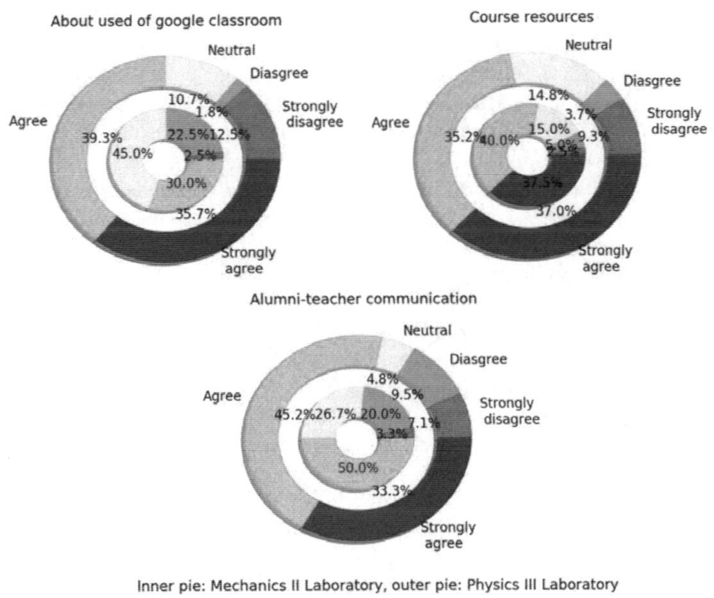

Figure 1. Results of Likert items application on physics III laboratory and mechanics II laboratory. The questions are organized by groups: about used of google classroom, course resources, and alumni-teacher communication. The colors mean Likert scale: strongly disagree (red), disagree (orange), normal (yellow), agree (green) and strongly agree.

The results of the Likert items showed that students receive very well the implementation of technological tools in the laboratory courses. Because they think it's a way to keep contact with the teacher outside the classroom, they keep informed in real time the flow of activities, including their grades. When they keep contact allows them to understand that they are in a process of collaborative learning instead of the classic learning process, the teacher as the transmitter and the student as the receiver. The students wait for mediated learning to be implemented in the theoretical courses with an online platform. They think that google classroom so much simple and friendly than others online platforms for teaching.

A teacher who wants to use the platforms for teaching could think to use the tool is more work to him. On the contrary, we affirm to keep course organized, makes teacher's work more efficient and then easier. It´s used to measure opinions, perceptions and response to a statement.

4. Conclusion

Strategies to strengthen the writing and learning processes from physics laboratories, such as elements of mediated learning combined with (JiTT) were implemented through the google classroom web service. The impact of the methodology proposed here was measured through Likert items, resulting in acceptance by the students of engineering and physics, as well as two orderly and reproducible courses (physics III laboratory and mechanics II laboratory) for the next academic periods.

Acknowledgements

We are grateful to the School of Physics at the Universidad Industrial de Santander in special to his Director: Jorge Martinez Téllez for his approbation in order to implement the web service google classroom at the laboratories for physics and engineers.

References

[1] MEN C 2009 *Deserción estudiantil en la educación superior colombiana* (Bogotá: Imprenta Nacional de Colombia) p 158
[2] Losio M and Macri A 2015 *Revista Latinoamericana de Políticas y Administración de la Educación* **3** 114–126
[3] Sanchez Torres F, Marquez Zúñiga J 2012 *La deserción en la Educación Superior en Colombia durante la primera década del siglo XXI: ¿por qué ha aumentado tanto?* (Bogotá: Universidad de los Andes)
[4] Tabarquino Muñoz R A 2015 *Estudios Políticos* **46** 221–241
[5] Amariles F, Paz O P, Russell N and Johnson N 2006 *The Journal of Community Informatics* **2(3)** (http://ci-journal.org/index.php/ciej/article/view/25 http://ci-journal.org/index.php/ciej/article/view/2566)
[6] MEN C 2017 Plan Nacional Decenal de Educación 2016-2026 *El camino hacia la calidad y la equidad* (Bogotá. Ministerio de Educación Colombiano) p 84
[7] Riascos-Erazo S C, Quintero-Calvache D M and Ávila Fajardo G P 2009 *Educación y educadores* **12** 133–157
[8] Miranda D A, Sanchez M J and Forero O M 2017 Laboratory preparation questionnaires as a tool for the implementation of the Just in Time Teaching in the Physics I laboratories: Research training *Journal of Physics: Conference Series* **850** 012015
[9] Adeyemo S A 2010 International *Journal of Educational Research and Technology* **1** 48–59
[10] Perkins K, Adams W, Dubson M, Finkelstein N, Reid S, Wieman C and LeMaster R 2006 *The Physics Teacher* **44** 18–23
[11] Chinnery G M 2006 *Language Learning & Technology* **10** 9–16
[12] Novak G M 2011 *New Directions for Teaching and Learning* **2011** 63–73
[13] González H L, Palencia A P, Umaña L A, Galindo L and M L A V 2008 *Advances in Physiology Education* **32** 312–316

IOP Conf. Series: Journal of Physics: Conf. Series **1161** (2019) 012021 doi:10.1088/1742-6596/1161/1/012021

Learning from the construction of a fidget spinner in a geometry classroom project

DA Prada[1], A Acevedo[2], D Serrano[3], MP Sanabria[1], F Duran[4] and J Gómez[5]
[1] Grupo de Investigación en Materiales, Universidad Pontificia Bolivariana, Bucaramanga, Colombia
[2] Grupo de Investigación en Administración, Universidad Pontificia Bolivariana, Bucaramanga, Colombia
[3] Grupo de Investigación Detección de Contaminantes y Remediación, Universidad Pontificia Bolivariana, Bucaramanga, Colombia
[4] Grupo de Investigación en Tecnología del Plasma, Universidad Pontificia Bolivariana, Bucaramanga, Colombia
[5] Centro de Gestión de Mercados, Logística y Tecnología de la Información, Servicio Nacional de Aprendizaje (SENA), Bogotá, Colombia

E-mail: duwamg.prada@upb.edu.co, david.serrano@upb.edu.co

Abstract. Knowledge development is a challenge for Faculty members. Understanding the classroom and societal dynamics allow to think other ways to learn. The fidget spinner is becoming popular among students. That is why in a geometry class from Universidad Pontificia Bolivariana (UPB), the final project was to construct this artefact. During the activity, abilities to design were strengthen as well as to build from scratch the spinner. Students comprehend how it works but also can create new alternatives to improve the spinner cycle. Concepts became clear and the function of geometry take a better relevance for student's life.

1. Introduction

During first undergraduate levels of any area of knowledge, it is common the need to address the concepts taught by the teacher, it helps to the professional contexts in which students will play once their university education is completed. Geometry and Trigonometry course is not an exception. For Engineering careers in general, education demands a great contribution in terms of applications, that is why here is proposed a classroom project that seek to bring the student to more practical ideas related with the subject contents. An example for Euclidean Geometry could be figures of an entire dimension that allows us to describe and create approximations of nature objects or human creation [1] [2]. Understanding these concepts, often turn out to be abstractions and the time intensity for the teacher mediation can be counterproductive for the relevance that wants to be granted. In order to achieve a better result, the use of a specialized Software has been proposed as an excellent ally to complement the accompaniment provided by the teacher. For this part, not only the use of a specific program such as GeoGebra is achieved, but also, it is possible to give a first approach of the Software that students will cover in their professional careers such as SolidWorks and AutoCAD.

The continuous advancement of technological tools, for example increasing specialized software and easy access allow to energize the curriculum in terms of self-knowledge and skills that are derived from the use of digital resources. After all, new technologies will end up eroding the curriculum and demanding adaptation and improvements, intimately articulated to these technologies [3]. This technological path of seeking to improve education does not imply leaving behind the use of traditional teaching tools that have provided significant learning [4]. It is important to recognize that the above demands Faculty a continuous reflection on their pedagogical work articulating the ICT and the student understanding of the concepts that are socialized in the classroom. That is because there may be shortcomings at the level of technological development, which show errors in the procedural part adjusted to an interpretation of low foundation. In general, this type of activity is evident at the moment of making a didactic transposition, understood as the process by which the action of transposing a knowledge towards a didactic site takes place [5] and not as an analogy of what is desired understand. These actions within the classroom adapt mathematical knowledge to transform it into knowledge to be taught [6], that is, transforming an acquired mathematical knowledge and bring it to an adaptation through technological means through a didactic intervention aware of limiting as much as possible the formation of wrong concepts [7]. The importance of generating a lasting knowledge is evidenced when new information is related to some aspect of what already exists in the cognitive structure of the individual, and thus a process that leads to significant learning is produced [8]. Understanding the different variables that affect a case study, allows to create spaces for analysis, discussion and interpretation that lead to real and significant concrete knowledge for those who handle it. Although equations have traditionally been used as recipe for problem solving, they provide a deeper insight when are used as a guide to think [9].

2. Methodologic project guide

Geometry and Trigonometry subject is oriented in the first academic semester of the engineering cycle at the Universidad Pontificia Bolivariana (UPB). A fidget spinner creation has been assigned as the Final Project of the classroom. Fidget spinners are increasing in popularity and, the use of it as a strategy to learn is emerging. The mechanism behind spinners is relatively simple. As any spinning apparatus, fidget spinners rotate around central axis, formed by two rings. By using a ball bearing mechanism instead of simple sliding between the rings, friction may be reduced significantly during rotation. In order to further increase the duration of the rotation, fidget spinners are equipped with three wings (for most spinners) bearing weight distributed equally from the center. [10] This type of artifact also has an impact on the level of improvements in the levels of attention and performance in people with some physical disorders. Fidget spinners are advertised to increase concentration and attention to academic tasks. Past research suggests that hyperactive movements, such as fidgeting, improve performance on attention tasks in children with attention-deficit / hyperactivity disorder (ADHD) [11]. Although also due to the great acceptance of this artifact, many of the students have used it more as a distractor, given the overwhelming interest, countless students started taking their fidget spinners to classrooms, prompting US and UK teachers to ban the toy in schools [12]. However, it is a commitment of the teacher, take those distractions from the environment and mold them in such a way that they can be used in the classroom, that is, teacher can use the toy as an ideal way to present some fundamental concepts of physics [13].

Through topics worked in the class, such as geometric properties of polygons and circles, area, perimeter and volume, students will create a prototype spinner fidget by using a software, starting from an equilateral triangle. In order to generate a document that relates the objectives, the competences to be evaluated and the construction parameters, a methodological guide has been socialized to delimit the activity scope.

2.1. Location

The present project was developed with students of the Universidad Pontificia Bolivariana who are studying the subject of Geometry and Trigonometry. For the same development, looking for transversally in each subject of the basic cycle, student must present basic knowledge of angles, triangles, circular movement, circles and the use of GeoGebra software, SolidWorks or other dynamic software.

IOP Conf. Series: Journal of Physics: Conf. Series **1161** (2019) 012021 doi:10.1088/1742-6596/1161/1/012021

2.2. Objectives

a) Analyze and argue the geometric characteristics of a spinner construction.
b) Contextualize concepts learned in Geometry and Trigonometry engineering subject.
c) Design and construct, through a geometric analysis, step by step of a spinner.

2.3. Justification

a) Students, through mathematical and geometrical concepts, will analyze the different characteristics of a spinner in terms of triangles, circles, arcs of circumference, area, perimeter and angular velocity.

2.4. Cognitive skills

a) Understand analysis parameters and geometry argumentation to give an explanation of the construction of an object.
b) Analyze the geometric structure necessary for the construction of a spinner.
c) Derive necessary formulas for the construction of a spinner using geometry as a tool.

2.5. Communicative competencies

a) Communicate your project in a fluid and coherent manner.
b) Prepares a report with coherence and cohesion, which reflects research results carried out on the project.

2.6. Procedural skills

a) Investigate seriously and give results as stipulated.
b) Analyze which are the most suitable design conditions for the object construction.
c) Understand the phenomenon behavior according to environmental variability.

2.7. Attitudinal competencies

a) Assess other members' points of view to teamwork.
b) Promotes teamwork.
c) It is consistent with the Bolivarian principles and values of the Pontifical Bolivarian University.

3. Construction method

A triangle of side l proportional to the measurement of the circumference of the bearing is proposed. This triangle must be equilateral (see Figure 1(a)). Then the mediatrices of the triangle are located with respect to each vertex (see Figure 1(b)), The point of the center is the equilibrium point of the figure, this will be the center of the bearing, then the bisectors of the triangle are located with respect to each vertex (see Figure 1(c)) where the center point is the equilibrium point of the figure, this will be the center of the bearing.

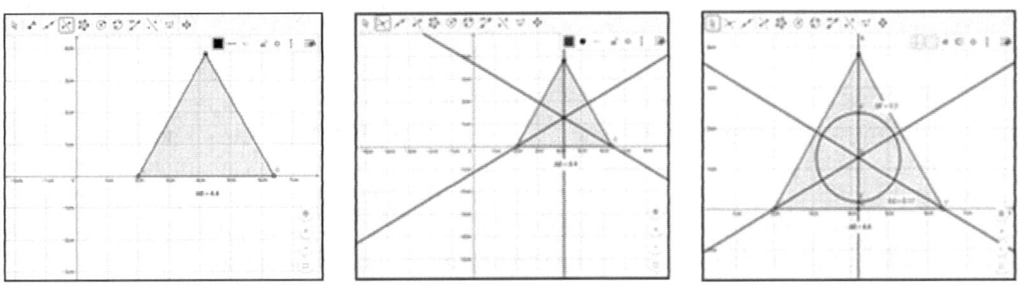

(a) (b) (c)

Figure 1. (a) Equilateral triangle, (1) mediatrix and (c) breakeven. Images made with Geogebra.

Next, a circle is drawn from the center previously found and with a radius that is half of the measurement of the diameter of the bearing (see Figure 2(a)). There are cut points of the perpendicular bisector with the side of the triangle and the circumference. The cut point on the side of the triangle is the center and the cut point with the circumference is the radius of the circumference that will form a crown with the circle previously inscribed (see Figure 2(b)). Finding the cut points of the perpendicular bisector with respect to the circumference and taking as center the upper vertex, two circles are made in the upper part of the triangle (see Figure 2(c)).

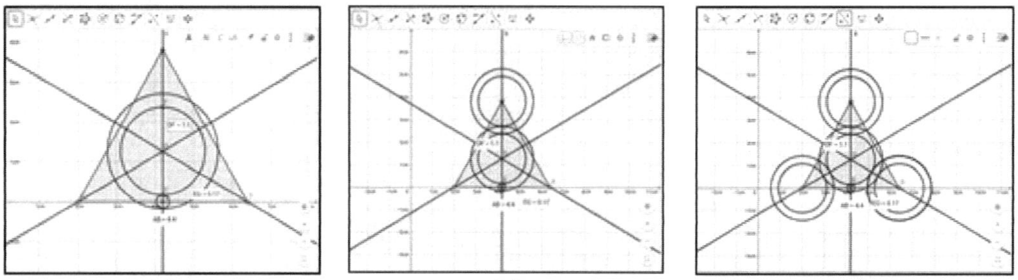

(a) (b) (c)

Figure 2. (a) Circle, (b) two circular crowns, and (c) four circular crowns. Images made with Geogebra.

By axial symmetry, circles previously made are copied, which represent the ends of the spinner (see Figure 3(a)), Likewise by axial symmetry the circles inscribed in the center of equilibrium of the triangle are copied which will help to form the arcs that give shape to the spinner (see Figure 3(b)). With cut points between circumferences and the perpendicular bisector, three-point arcs are formed on each side of the spinner (see Figure 3(c)).

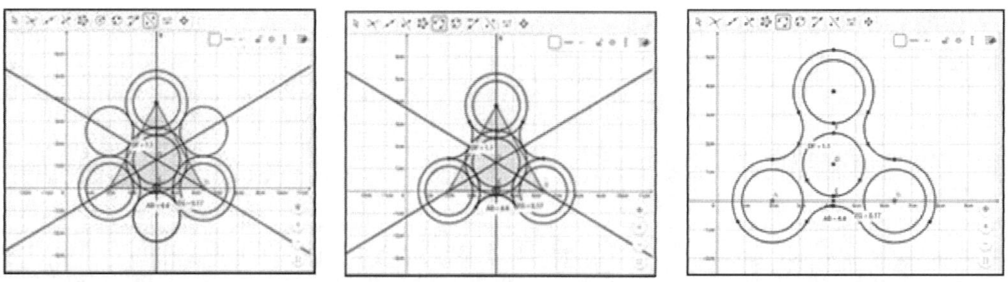

(a) (b) (c)

Figure 3. (a) Axial symmetry, (b) axial symmetry, and (c) arches. Images made with Geogebra.

Finally, to design details, objects that are not necessary are hidden to better appreciate the spinner design.

4. Results and discussion

After design making and spinner construction, it was found that weight is a factor that changes the way it rotates. The stability depends on the proper distribution of weight in each of its spinners. On the other hand, the center of gravity should always be in the center circle to avoid sideways movements. The materials of the construction can vary, according to the possibilities of obtaining them. In the class classroom activity, the spinners rotated on average 3 minutes. Figure 4 shows the two spinners that had the most time in rotation.

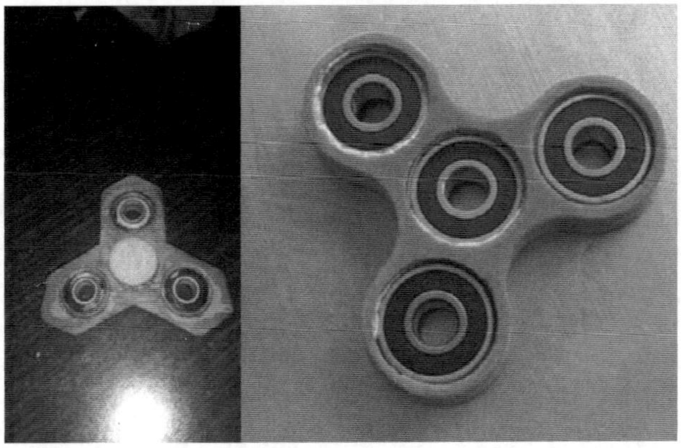

Figure 4. Spinner built by students.

5. Conclusion

From the activity it was possible to observe the use of the different topics worked in Geometry class regarding the construction of the spinner. The students argued that the rotation time depends on the material used and the weight at the ends of the spinner, since it can affect its angular velocity. The laser machine generated a margin of error in the measurement of the bearing, therefore the space between the bearings and the spinner mold had to be filled.

Due to the use of Euclidean Geometry, students were able to come up with different designs, opening the experience to the achievement of different prototypes that managed to develop creativity and change the conception of what the mathematical and technological methodology provides them. In addition, the construction itself allowed to address the analysis of physical characteristics that optimized or worsened the functionality of the Spinner, as was the incidence of the weight of the ends in the angular velocity, as well as the accuracy in cutting the main body, which would allow to balance its weight in each of its parts and thus achieve a stable turn and, therefore, more sustained.

References

[1] Mandelbrot B and Wallis J 1969 *Wat. Res. Res.* **5** 321-340
[2] Mandelbrot B and Wallis J 1969 *Wat. Res. Res.* **5** 967-988
[3] Moreno L 2014 *Educación matemática: del signo al píxel* (Bucaramanga: Universidad Industrial de Santander)
[4] Andrade H and Gómez L 1997 *Tecnología informática en la Escuela* (Bucaramanga: Universidad Industrial de Santander)
[5] Cantoral R and Farfán R 2004 *Desarrollo conceptual del cálculo* (Colombia: Thomson Learning)
[6] Chevallard Y 1985 *Rev. Fran. Péd.* **76** 89-91
[7] D'Amore B 2011 *Didáctica de la matemática* (Colombia: Cooperativa Editorial Magisterio)
[8] Novak J 2000 *Aprendiendo a aprender* (Barcelona: Reverté) pp 220-343
[9] Hewitt P 2017 *Sci. Teach.* **84(3)** 14-18
[10] Cohen E, Bravi R, and Minciacchi D 2018 *Sci. Rep.* **8(1)** 3144
[11] Sarver D, Rapport M, Kofler M 2015 *J. Abn. Child. Psy.* **43** 1219–1232
[12] Strauss V 2017 *Schools are banning fidget spinners, calling them nuisances and even dangerous* Consulted on: https://www.washingtonpost.com/news/answer-sheet/wp/2017/06/01/schools-are-banning-fidget-spinners-calling-them-nuisances-and-even-dangerous/?noredirect=on
[13] Lindén J 2018 *Phys. Educ.* **53** 023004

IOP Conf. Series: Journal of Physics: Conf. Series **1161** (2019) 012022 doi:10.1088/1742-6596/1161/1/012022

Measurement of the performance of the inverted classroom methodology in the finance learning environment: A comparison with the traditional class

I Aznar[1], D A Prada[2], A Acevedo[2], F Durán-Flórez[2] and J Gómez[3]
[1] DOE, Universidad de Granada, Granada, España
[2] GIA, Universidad Pontificia Bolivariana, Bucaramanga, Colombia
[3] Centro de Gestión de Mercados, Logística y Tecnología de la Información, Servicio Nacional de Aprendizaje (SENA), Bogotá, Colombia

E-mail: iaznar@ugr.es, duwamg.prada@upb.edu.co

Abstract. The inverted class or Flipped Classroom teaching methodology that proposes that the work of certain learning processes be transferred out of the classroom, that is to say that the students prepare the subjects outside the alma mater, through the use of information and communication technologies (TIC) and uses time in the classroom to carry out activities that require greater participation and student - teacher interaction, to debate ideas and development of constructs and mindfacts appropriate to generational change, management that is sustained with the support of new technologies; In conclusion, the method under study reverses the activities that are normally done in classes, with the activities that are done at home, that is, the technique proposes a change of paradigm more adjusted to the current realities of teaching. In order to demonstrate the potential of the Inverted Class proposal, the professors participating in this research reviewed in depth the results achieved with the application of this technique within the subject Capital Market pertaining to the degree of Business Administration and the financial area specifically to compare the results; To obtain the corresponding metrics, a perception measurement was displayed to students of two cohorts for the year 2017, distributed in 4 rooms per semester, and an average of 25 students per class, for a total of 212 students participating in the methodology belonging to the schools of economic and accounting administrative sciences of the private Universities of the region, where the technique object of study was applied and orienting others under the traditional system of education. This was elaborated in light of the case research approach, abstracting the results through the use of qualitative and quantitative methods, obtaining empirical evidence in the significant improvement of the students in terms of results and appropriation of competences, with an increase in the participation of students, and a strong orientation with discipline to the management of tasks and the research component, through interaction with technological tools placed at the service of pedagogy.

1. Introduction

With the technological advances achieved by society, it is a place that does not have an education can't be excluded, which is why it must be adjusted to the changes that day by day in the world in life, elements such as information Communication and work styles have been changes in the transformative form for the 21st century [1]. By linking technology to learning processes, the symbiosis that can be used to take advantage of learning environments. The relationship between technology and the teaching-learning processes is not to share the new one, and these have been evolving and strengthening the academic processes.

The inverted class has been used as a strategy in the experiences developed within subjects of the branch of economic sciences [2], through the use of technology as support in the learning processes in the classroom. Another thing that can be rescued in the genesis of methodologies supported by technology was developed by North American chemistry professors from Woodland Park School in Colorado, to whom the scientific community is endorsed by the consolidation of the term Inverted Classroom [3]. This is a pedagogical approach in which "direct instruction moves from the collective learning space to the individual learning space, and the resulting space is transformed into a dynamic and interactive learning environment in which the educator guides the students learn in the subject "[4], it is also indicated that the technique is about" a didactic model in which students learn new content through online video tutorials, usually at home [5] what used to be "duties"(assigned tasks), are now carried out in the classroom with the teacher, the tracking of the evidences, the samples of the technology of the technological tools, the teaching process, the teachers, the help, the task of the students, the learning processes, the best assignment of competences to achieve the dynamics of this methodology, it is necessary, in addition, of a student, in an efficient way, a commitment, and in any other way, be the builder of their own learning, what can be introjected and then what is multiplied and / or integrated into their context or reality. Likewise, [6,7] agree that the inverted classroom technique considers the delivery of content including video conferences, to be reviewed outside the classroom, focusing class time in the development of exercises and applications in class, and that from his observations is an innovative method to create improvements in student learning.

In order to achieve the above, the commitment of all those involved in the educational process is necessary, demanding from them a competence not only in what is particularly socialized in the classroom, but also competence in the use of technological tools such as multimedia resources of learning, Virtual Learning Objects (OVAS), online discussion forums, and other instruments and tools available for this purpose that support said strategy; added to this is the development of strategies, construction and preparation of material aligned to what is necessary for the student, then this methodology reverses traditional models in the teaching-learning process, facilitating online work patterns from outside the classroom and moving certain tasks within the class [8].

There is little research done in the field of finance regarding the teaching methodology to be developed with the students and even more, when we have ICTs in mind to favor an approach of the contents in a more current way, so the Flipped Classroom or inverted class we consider it an adequate method to favor this learning and the results of research with this form of learning made in other areas of knowledge indicate that (see Table 1):

Table 1: Results of research on the Flipped classroom methodology.

Research	Results
Opazo, Acuña y Rojas (2016)	Increase of 5% of grades, greater autonomy in learning, more participation and solving doubts
Díaz, Martín y Sánchez (2017)	Higher learning outcomes, increased student motivation, acquisition of generic and specific skills
Hinojo, Mingorance, Trujillo, Aznar y Cáceres (2018)	Greater performance, more involvement in the subject and, consequently, less university dropout.

In [9] the impact of the flipped classroom methodology on the learning outcomes of (28) students was evaluated, comparing them with the previous cohort of (56) students who had experienced a traditional methodology. Using the application with a pretest and posttest measure, the learning gain was calculated observing the results shown in Table 1. Another investigation in which the efficiency of the flipped classroom methodology was observed was developed in the discipline of Operations Management, where a control group (136 students) and an experimental group (65 students) were taken applying a pretest and posttest to compare both groups. The results obtained also reflected the importance of using this tool [10]. The research developed by Hinojo, Mingorance, Trujillo, Aznar and

Cáceres, aimed to analyze if there are significant differences in academic performance and if attendance improves, depending on the methodology used in the overall teaching-learning process, with the paradigm of the intervention change, in which the main axis is centered on the student, that is, to know if the flipped classroom methodology improved academic performance [11]. For this study, an empirical-analytical research was implemented, with quasi-experimental groups of two groups of future physical education teachers, with a sample of 131 students. Finally, after the implementation of the different sessions, it is found in the results that the experimental group increased its ratings with respect to those that followed a traditional methodology.

For all the above, we can define a definition of Flipped Classroom or inverted class as an active learning method centered on the student, which allows the teacher to turn the classroom into a place to solve problems, advance concepts and participate in collaborative learning [10].

Exposed the spectrum and the implication of the formative process, and the use of technology, so that said process is more in accord with the times of today, the present investigation was oriented to investigate and to specify the perception of the university students of the city of Bucaramanga about the experience and / or experience they have had with the subject of the inverted classroom, an activity that was developed to greatly facilitate the appropriation of knowledge online, based on projects for the Capital Market subject, taught at Universidad Pontificia Bolivariana (UPB) and Universidad Santo Tomás de Aquino (USTA) of the city of Bucaramanga, Colombia.

The proposal made the main objective of improving the quality of teaching and the efficiency of learning, to deploy and observe the performance in each of the two semesters of the year 2017. For this research, there was a participation of 402 students and 212 of they were instructed under the inverted class methodology. To observe the results of this work, mixed research methods were used, instruments necessary to explain and understand the perception of students in reference to the teaching strategy under study. The results show a significant improvement in students' learning processes, coupled with motivation and interest in learning (attitude), which undoubtedly leads to collaborative work that multiplies in the teaching-learning processes. Finally, invitations related to research and classroom methodology are raised.

2. Methodology

Each semester, the Capital Market class (which has 3 credits) hosts an average of 26 students, divided into 2 groups per semester in each university (4 groups per semester between the two institutions) which must be attended by the orientation teacher of the subject. The course presents requirements for the approval of other subjects in the area of finance necessary for the development of the programmatic content of Capital Markets. The time of interaction in the classroom is 16 weeks, in which two competency checks are made by means of a partial exam and follow-up notes such as quiz and / or short or punctual controls of a specific topic and the use of cases of study for the learner to present his / her opinions with reference to the topic addressed.

The inverted classroom is feedback with ICT considering it a methodology based on them, since mainly their previous work is supported in the observation of short videos, basic presentations of the thematic axis to be treated. Through the use of OVA's fourteen products were prepared, recorded and edited with the content of the main theories necessary for teacher mediation. The videos have a time between 10 and 35 minutes, with an average of around 14 minutes. These support tools, with the thematic axes described in the curricular syllabus, are shared and consulted by the students, in the EDOOME free educational platform five days before the face-to-face activity, with the aim of generating a reflection space that potentializes the interaction of the student based mainly on the use of MS-Excel spreadsheets.

Teacher mediation begins with a short feedback of the thematic axis that has been socialized in the videos and the supporting material, and then one or several questions are shared for the students to address during the development of the class, generating enough time to clarify the doubts that arise in the self-taught, which allows to explore in depth the issues and objectives raised in the development of the chair. After granting a period of no more than 15 minutes in the preparation of the assignment, the results obtained are discussed and the objective of teaching the class proper is discussed. It is fundamental that the self-learning component of the students is evidenced by the student's interaction based on their prior learning related to the virtual learning object that has been socialized, the tracking of these videos on this platform can't be controlled since only Record the download of the video only once.

3. Results

For this study, a quasi-experimental design was implemented in two groups of students of the capital markets subject. A control group of 190 students who followed the traditional methodology and the experimental group 212 students who followed the Flipped classroom. To obtain the results, the techniques of the eclectic method of research were used, which aims to show the progress made by the students addressed with this dynamic, comparing two teaching environments traditional and inverted class. The evidences obtained are shown under the quasi-experimental structure implemented for the study, since of the four groups addressed per semester, two of them were applied, during 16 weeks of academic work in each semester of 2017, the treatment condition while at the other two do not.

To minimize the biases given by the quasi-experimental technique, rooted in the issue of randomness of the units and treatments, the activity is carried out using all the thematic axes described in the syllabus curricula, and in the same way the evaluation instruments as it was quiz, case of studies and partial exams, given in two moments of each academic semester. With these instruments it was possible to assess the efficiency and effectiveness of the inverted class method by observing the cumulative results achieved; these quantitative metrics were potentiated with qualitative metrics obtained from the answers provided in the completion of the structured tool questionnaire for this purpose.

The evidences were observed through scales such as student cohesion, innovation and academic performance. These three domains were analyzed by means of an instrument applied electronically and anonymously, which yielded a p. valued ($p<0.05$). The consistency of the instrument is validated with the alpha coefficient of Crom Bach. The 212 participating students completed the survey where each respondent considered the three stages indicated above to provide their corresponding assessments and answers. The general reading of the exercise is considered to be well aligned with the quantitative results. The survey consists of a series of different questions about the course under study, the general learning environment, etc.

To validate the scope of the methodology, the application of a questionnaire was considered in order to obtain the insights and experiences of the participating students with the methodology proposed for this purpose. The questionnaire was answered by 212 participants (men: 57%, age: 20.6±4.13 years and 43% women, age 21.5±4.10 years). The overall score of the exercise and that of its variables were very outstanding with the inverted classroom strategy, with which the technique shows the positive impact of the same in this subject (see Table 2).

Table 2. Observed and expected frequencies regarding the updated contents.

Calcification	Frequencies Observed			Expected Frequencies		
	Students	Professional	Total	Students	Professional	Total
1 - Totally disagree	0	4	4	0.57	3.43	4
2 - In disagreement	2	7	9	1.27	7.73	14
3 - Agree	20	117	137	19.39	117.61	99
4 - Totally agree	8	54	62	8.77	182	95
Total	30	182	212	30	182	212

With respect to statistical inference, a non-parametric statistical analysis procedure was applied, such as the Pearson Chi-square test, for independent samples. The purpose of using this statistic was of statistically significant differences between the students participating in the measurement of the technique and the traditional model in terms of the scale of self-perception of initial learning. To measure the statistically significant differences, a level of confidence ($1-\alpha$) in 95% and alpha (α) of 5% were considered (see Table 3). For the subject of innovation, the question evaluated was: do you consider that the contents to be studied may be more current if the ICT resources are used as support for the methodology followed in class?

Table 3. Statistics used.

Calculated Chi	Tabulated Chi	Degrees of Lib.	P - value	Valid Cases
0.000	7.815	3	0.000	212

On the other hand, to clarify the students' self-perception regarding the contents, the following question was asked: Do you consider that a change in the traditional methodology can influence your academic performance? Applying the same statistical process, with a level of significance of 5% (0.05), which resulted, as a result, a test statistic Chi^2=0.00, that is, that the hypothesis that there are significant differences between the students linked to the investigative process. It is concluded that there is no difference between the opinions of the groups studied, the former obtains its virtue in the result of the statistic p-value=0.000, evidence that the Ha, is rejected (see Table 4 and Table 5).

Table 4. Perception of methodology and the impact on academic performance.

Calcification	Frequencies Observed			Expected Frequencies		
	Students	Professional	Total	Students	Professional	Total
1 - Totally disagree	0	4	4	0.57	3.43	4
2 - In disagreement	2	12	14	1.98	12.02	14
3 - Agree	9	90	9	14.01	84.99	99
4 - Totally agree	19	76	95	13.44	81.56	95
Total	30	182	212	30	182	212

Table 5. Statistics used.

Calculated Chi	Tabulated Chi	Degrees of Lib.	P - value	Valid Cases
0.000	7.815	3	0.000	212

For the third stage under analysis, the following question is asked: Do you consider that a change in the traditional methodology can influence your content retention, being this better and more lasting? Like the previous two, a level of significance of 5% (0.05) was taken, resulting in a test statistic Chi^2= 0.00, that is, the hypothesis that there are significant differences between the groups is rejected. object of study. It is concluded that there is no difference between the opinions of both students with the inverted class treatment and students oriented with traditional methodology, the above is demonstrated by the revision of the p-value which yields a result of 0.000, evidence that the Ha, it is rejected (see Table 6 and Table 7).

Table 6. Change of methodology and content retention.

Calcification	Frequencies Observed			Expected Frequencies		
	Students	Professional	Total	Students	Professional	Total
1 - Totally disagree	0	4	4	0.57	3.43	4
2 - In disagreement	2	16	18	2.55	15.45	18
3 - Agree	14	91	105	14.86	90.14	105
4 - Totally agree	14	71	85	12.03	72.97	85
Total	30	182	212	30	182	212

Table 7. Statistics used

Calculated Chi	Tabulated Chi	Degrees of Lib.	P - value	Valid Cases
0.000	7.815	3	0.000	212

IOP Conf. Series: Journal of Physics: Conf. Series **1161** (2019) 012022 doi:10.1088/1742-6596/1161/1/012022

Regarding the perception of the change of methodology and content retention, developed this stage with a level of significance of 5% (0.05), delivered as a result, a test statistic Chi2 =0.00 that is, rejects the hypothesis that there are significant differences between the study groups. It is concluded that there is no difference between the opinions of both graduates and employees; a p-value =0.000, evidence that the Ha, is rejected.

In addition to the process of statistical inference, and to demonstrate the benefits that arose from the application of the methodology of the class invested in the development of the Capital Market subject, in Table 8 and Table 9 the notes reached in each of the participating groups, which is presented below,

Table 8. Numerical results of the tests applied to the capital market matter first half 2017.

Measuring Instrument		Inverted class students			Students traditional method		
	State	Q student	%	Average score	Q student	%	Average score
Quiz total	Approved Reprobate	48 55 103	47 53	4.2 2.8	32 73 105	30 70	3.9 2.5
Study cases total	Approved Reprobate	49 54 103	48 52	4.4 2.7	45 60 105	43 57	4.1 2.6
1st cut evaluation total	Approved Reprobate	50 53 103	49 51	4.0 2.7	40 65 105	38 62	3.6 2.2
2nd court evaluation total	Approved Reprobate	51 52 103	50 50	4.1 2.8	41 64 105	39 61	3.7 2.3

Table 9. Numerical results of the tests applied to the capital market matter second semester 2017.

Measuring Instrument		Inverted class students			Students traditional method		
	State	Q student	%	Average score	Q student	%	Average score
Quiz total	Approved Reprobate	47 62 109	43 57	4.3 2.8	33 52 85	39 61	3.8 2.5
Study cases Total	Approved Reprobate	47 62 109	43 57	4.4 2.8	46 39 85	54 46	4.2 2.8
1st cut evaluation total	Approved Reprobate	47 62 109	43 57	4.4 2.8	39 46 85	46 54	3.8 2.3
2nd court evaluation total	Approved Reprobate	49 60 109	45 55	4.5 2.8	38 47 85	45 55	3.8 2.4

The previous table shows a significant improvement with the class methodology invested in each and every one of the instruments used for the measurement of competencies, where the rating of 5.0 would be the highest rating and 0.0 the lowest, the data have relative consistency by those found in [12] and [10].

IOP Conf. Series: Journal of Physics: Conf. Series **1161** (2019) 012022 doi:10.1088/1742-6596/1161/1/012022

4. Conclusions

Among the most representative findings, it is evident that in the results obtained by the students of the inverted class, the grades obtained in the four measurement stages (quiz, jobs and exams) are higher in reference to the students assisted by the traditional class. Now, in terms of the perception that students apply in three scenarios, it is evident that there are no significant disagreement results among the perceptions of the groups studied, arriving to a large extent to say that the application of this technique brings significant learning for the potential professionals.

References

[1] Niess M 2005 *J. Edu. Technol. Soc.* **21(5)** 89-100
[2] Lage M, Platt G and Treglia M 2000 *J. Eco. Edu.* **31(1)** 30-43
[3] Bergmann J and Sams A 2012 *Flip your classroom: Reach every student in every class every day* (Washington: ISTE)
[4] Bergmann J and Sams A 2014 *Training & Development* **68(2)** 28-31
[5] Tourón J and Santiago R 2015 *Rev. Edu.* **368** 196-231
[6] Deslauriers L, Schelew E and Wieman C 2011 *Science* **332(6031)** 862–864
[7] Mason G, Shuman T and Cook K 2013 *IEEE Trans. on Edu.* **56(4)** 430-435
[8] Barreras M 2016 *Educatio Siglo XXI* **34(1)** 173-196
[9] Opazo A, Acuña J and Rojas M 2016 *Innoeduca* **2(2)** 90-99
[10] Díaz E, Martín M and Sánchez J 2017 *Working Papers on Operations Management* **8(S.I.)** 15-18
[11] Hinojo F, Mingorance A, Trujillo J, Aznar I and Cáceres M 2018 *Sustainability* **10(5)** 1334
[12] Baepler P, Walker J, and Driessen M, 2014 *Computers & Education* **78** 227-236

IOP Conf. Series: Journal of Physics: Conf. Series **1161** (2019) 012023 doi:10.1088/1742-6596/1161/1/012023

TelEduc: A virtual learning environment for teaching and learning at the University of Campinas, Brazil

J P Acevedo-Rincón[1] and C E Flórez-Pabón[2]
[1] Edumat, Universidad Industrial de Santander, Bucaramanga, Colombia
[2] Conquiro, Universidad de Pamplona, Pamplona, Colombia

E-mail: jepaceri@uis.edu.co, ceflorez@unipamplona.edu.co

Abstract. The present paper aims to describe the learning experiences of some students during their pre-service stage as future teachers between 2015 and 2017. Approximately 80 students have participated in the development of this proposal. These students belong to three different programs: Pedagogy in Early Childhood Education, Bachelor of Mathematics Education, Bachelor in Physics and Chemistry from the State University of Campinas, Brazil. As part of their practice they use the teaching and learning system called Open-Teaching, which is available for undergraduate courses at Unicamp. This system automates the registration of courses in the TelEduc and Moodle environments. Both Virtual Learning Environments offer content and communication tools, allowing activities among the participants. TelEduc supports the teaching learning processes for the three education programs including exact sciences and pedagogy.

1. Introduction

Virtual Learning Environments (VLE) have become an option to support teaching at different levels of Education. This is, new technologies are used with a learning purpose becoming "a relevant input for the transfer of information and training, which allows a personalized process of tutoring, constant access to the program of the course, increased communication between students and tutors through the use of tools" [1]. Particularly, in higher education this type of environment responds to the teaching needs of university centers. In teacher education, in different knowledge areas, the use of these environments invites future teachers to experience other classroom environments. Developed countries have journeyed this road since approximately twenty years ago [2]. However, in countries with less possibilities, the inclusion of this type of environment in teacher training has been but with slow to incorporation. Despite of this, being able to offer this opportunity to future teachers, that of being in unconventional environments of education, allows them to develop technological skills, and possibly, when they are teachers in practice, use other favorable environments for teaching their knowledge areas. In this sense, the integration of VLEs to the teaching and learning of different sciences allows "altering the way educators go about their routines of delivering subject content" [3-4].

Different perspectives highlight the contributions of the inclusion of VLE as a means of interaction and relationship between teaching and learning in virtual education, such as WebCT (Web Course Tools) or Blackboard. However, other research leads to consider it as support to face-to-face education. Specialized companies from different geographical locations offer alternatives to the classic web tool offerings, such as the TelEduc, an e-learning environment for the creation, participation and administration of courses on the Web.

IOP Conf. Series: Journal of Physics: Conf. Series **1161** (2019) 012023 doi:10.1088/1742-6596/1161/1/012023

On the other hand, in relation to learning based on the use of these virtual environments, the theories of social learning, such as communities of practice [5], highlight the collaborative learning mediated by participation and reifications tools contained in the virtual environment, for example, by using elements such as asynchronous discussion forums and synchronous chat tools or on-line assessment [2] In this sense, in this document, we intend to present the characteristics of the VLE (TelEduc), which is currently used to support face-to-face education at Unicamp (Brazil). Subsequently, the description of the three experiences developed with regular courses at the University of Campinas, with groups of prospective teachers of the careers of Pedagogy in Early Childhood Education, Integrated Pedagogy degree in chemistry and physical, and mathematics education degree. From this, some advantages and disadvantages of the use of this VLE in the teaching and learning of prospective teacher as a result of analysis of these three experiences are presented. Finally, the conclusions about the comparison of the results of the three courses are narrated.

2. The VLE in the Unicamp

The State University of Campinas (Unicamp) in Brazil, is a public university in the state of Sao Paulo, which was founded in 1962. It currently offers courses in 70 graduation programs and 157 postgraduate programs, in addition to the coverage offered by the extension programs. The academic programs both face-to-face and extension (some distance). Some courses at Unicamp propose as a tool to support teaching, the TelEduc platform or ensino aberto (in Portuguese means Open teaching or Open university) as it is called in its original language (Portuguese).

The site hosted at https://www.ggte.unicamp.br/ea presents two input options, one for TelEduc, and one for Moodle. The teacher who runs the course decides which one to use, or if he needs the combination of the two. For the particular case of this article, we present the VLE called TelEduc, and its use in courses of the Faculty of Education of Unicamp.

Similarly, the TelEduc offers activities tools, such as agenda, readings, support material, frequently asked questions, groups, which store documents for consultation or work of students. Also, the VLE has as communication tools among the participants (students, teacher and / or invited teacher) the email, chat, forums, wall, portfolio and participant profile, for the participation and reification in Wenger's sense [5]. Other tools are offered by the environment for evaluation (Exercises) and monitoring of student performance (Assessment).

The VLE has evolved, to the point of adding to or System tools since version 3.8, as assessment and Exercises. The first one, allows the teacher to accompany the performance of the students. The second one, allows only to offer students questions and questionnaires. Currently version 4.4.0 of this VLE works, which allows to consolidate in a single user all the courses registered by the student, and online access. This is a light, free and open e-learning system, created in Brazil in 1997 as a result of a master's thesis of a Brazilian student, for classroom and distance education at a national and international level.

The contents stored in a course registered by the student remain in time, which can be accessed by both the teacher, teachers and guests. Its access and navigation is intuitive, and easy to access online. The TelEduc site, properly from its beginning, is in the web address as http://www.teleduc.org.br/. However, each university houses this site within the institutional, as in the case of Unicamp, which is today http://webensino.unicamp.br/pagina_inicial/index.php.

On the other hand, we can demonstrate multiple teaching and learning experiences based on the use of the TelEduc VLE. In this particular, the faculties and institutes adopt their own methodologies that involve the use of technological tools that facilitate both teaching support, as well as access to materials, proposals, exercises, and to communication tools such as forums and mail. The Faculty of Education promotes the use of TelEduc, since its inception, this product being a graduate of the master's program in education this faculty since 1997. Within the courses offered by the faculty and accompanied by the first author- The researcher of this article, as invited teacher, are Pedagogy in Early Childhood Education Course, Pedagogical Practices in Mathematics course and Physics and Chemistry Seminar. That is to say, three experiences lived with these regular courses at the University of Campinas will be narrated, in three different semesters, from the description of the contexts in which these courses were developed, their participants and the learning of the participants, in terms of using tools to analyze their practices as future teachers.

2.1. Pedagogy in early childhood education course

Within the education of prospective teachers of pedagogy, they have Pedagogy in Early Childhood Education, and internship stage in Early Childhood Education course. In the first semester of 2015, the first author was invited to participate in the both courses. With forty eight participating students, one main teacher and one visiting professor, the Pedagogy in Early Childhood Education course was developed during four horas per week, focusing on the use of communication tools, through forums, e-mail and the use of individual student portfolios, in which, they stored each one of the artistic registers proposed by them as evidence of learning and exploration of multiple languages in early childhood education [6], but under the exploration of the multiple languages of the prospective teacher of Early Childhood Education . With a total of eleven registered activities, in the individual portfolios, the face-to-face classes were developed under the perspective of debate with the readings of the support material, and at other times under the focus of case studies from the practices of education childish.

On the other hand, during the same semester, the internship stage in Early Childhood Education was the place where the students made their observations and accompaniment to the kindergarten of the region, in four hours at the week, as the beginning of their practices as a prospective teacher of Early Childhood Education. Therefore, it was broadly related to the first course offered, already mentioned in the previous paragraph, so, in the VLE of this course, it was only focused on the narrative record of its accompaniments, through the use of individual portfolios of the twenty-seven prospective teacher participants. From the use of these narrative records the prospective teachers managed to create an effective communication channel between them and the professors (holder and guest), since, for each portfolio uploaded by the student, a feedback of what was written was recorded by their teachers. In addition to this, some of the cases narrated in the portfolios were selected as case studies in the pedagogy course of Early Childhood Education. At the end, both the first and the second course of prospective teachers of pedagogy, the participants prepared a final report, in which elements were exposed in their respective portfolios, during the semester, which served to bring the evidences of learning from them.

2.2. The pedagogical practices in mathematics course

One year after the first the accompaniment of the first group, was developed the second experience. This is, in the second semester of 2016, twenty-seven students from Bachelor of Mathematics education, comes to the Education institute to participate in the Pedagogical Practices in Mathematics course. The course was developed with an hourly intensity of two hours per week, for sixteen weeks. The course teaches under the formative process, the learning and the argue about the mathematics teaching and learning practice in the initial formation for prospective teachers in the Pedagogical Practices in Mathematics Course. The course aim was to know and problematize the teaching and learning practices in the school. The activities of the prospective teachers were developed under the methodology Lesson Study (LS). Also, within the Figure 1 and in pink, there are the moments attended by each of the six groups, which correspond to the specific steps of the original Japanese LS: Goal Setting, Lesson Planning, Research Lesson, and Reflection and Systematization.

Prospective teacher developed a 'Glocal' Lesson Study from the choice of topic - relevant to the school curriculum - along with lesson planning, sharing and discussion of lesson proposals, lesson implementation, lesson analysis, and presentation/discussion of results, culminating in the writing of articles. To this study was very important the Teleduc VLE, because, each advance within the stages of the Lesson Study, was stored in the VLE, and by this means a first feedback filter of the class proposals was made. Also, the TelEduc portfolio, was a virtual space to store articles, textbooks and theses, with which the students could sustain their proposals for planning, application and reflection over the application. Below, we describe briefly some reactions and reflections from the students over each part of the LS cycle in the PPM Course as appear in Table 1, and the relationship with the actions in the TelEduc.

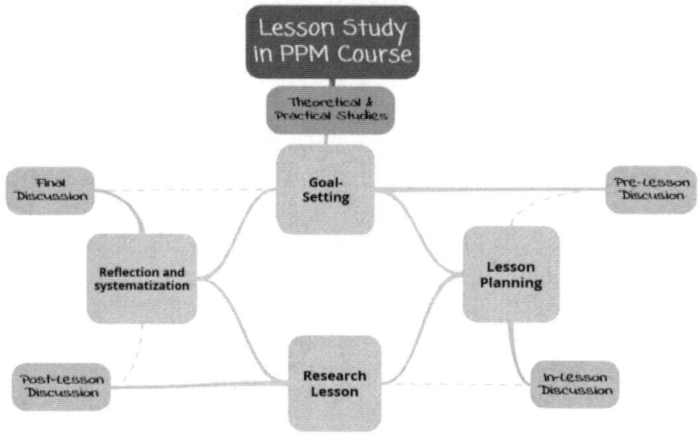

Figure 1. The process of lesson study in the pedagogical practices in mathematics course [7].

Table 1. Reactions and/or reflections from the students in the processes and sub-processes of Lesson Study Cycle.

Process in LS	Some reactions and/or reflections from the students	TelEduc Actions
Lesson Planning	The preparation of the lesson (Planning) is of great importance. Time and research should be devoted to thinking about a task / activity that fits the proposal being taught and thinking about the possible reaction of the students.	• Learning memorials (individual Portfolios). • Bibliographical references in Material of support in the TelEduc. • Proposal records of the lesson planning and feedback on each proposal on the group portfolios.
Research Lesson & In-Lesson Discussion	Discussing among us the information obtained in the discussions, execution and final result of the activity carried out, we conclude that exploratory-investigative classes contribute richly to the formation of the students.	• Observation and analyze records before the research lesson planning to record in the group portfolio.
Pos-Lesson Discussion	The activity objective was reached in the school since the students got involved in the search of the relations that we were proposing. The activities and approaches about the practices used are very efficient means to enrich professional development of graduate students.	• Discussing and record on the actions on the reflexive diary on group portfolios.
Reflections (systematization)	The activity proved to be a strong base for the students, because when we present the results we obtained for the Pedagogical Practices in Mathematics, and discuss these with the class, we concluded that we should go back to class and present other experiments to students.	• Reflections record in the portfolios. • In some cases, the students made a new proposal for the class, to start the LS cycle and saved in the group portfolio.
Final Discussion	The Lesson Study also gives us contribution when proposes a bigger planning and analysis of activities, an investigative posture of your own practices and the collaborative construction of activities.	• Record of the experience in article format, in order to share with other groups from the record and storage in group portfolio

We will highlight the process and some results of the implementation of the Lesson Study in a pedagogical discipline of the degree course in mathematics. The mail, the forums, and the portfolios were other tools of the VLE, which helped the communication and the debate about the proposals registered in the TelEduc.

2.3. The physics and chemistry seminar

One group from the Integrated Pedagogy degree in chemistry and physical participated in the chemical and physical Seminar in the first semester of the 2017. In this seminar participated thirty-three students, one teacher and one invited teacher. the course structure was based on: (i) Exhibition classes, based on selected bibliography, according to the content. (ii) Seminaries with invited professionals; and, (iii) Two assessments (Exam one and Exam two) plus participation/frequency score. In this course, the use of TelEduc tools was limited to the communication (mail), the course dynamics, and Material of support the portfolios. The students participated in all the seminar sessions, and the students participated in all the seminar sessions, and for each last class, they had to review the texts kept by the teachers in the Teleduc support material space. After consulting the material, the students participated in classes, along with the teachers. Doubts and requests about the passing of the subject, the students communicated through the VLE mail, where the teachers followed up. The course was very productive, so the students identified other forms of interaction in class and outside it, by bringing elements of the VLE, such as bibliographical references, and sharing them in the reflections raised by the teachers.

Up to now, three different experiences have been narrated as invited professor, with regular courses of undergraduate students at the University of Campinas, in three different time spaces. From the description of the contexts in which these courses were developed, their participants and the learning of the participants, in terms of using tools to analyze their practices as future teachers. In the three courses accompanied the professors tended for the effective use of the VLE called TelEduc, in which they made use of their tools, and the registration of student learning was strengthened, in the form of a narrative diary, a partial report, a final report or as a learning memorial.

3. The VLEs for the prospective teacher learning

As trainers of future teachers, we must be aware of the generational gap we have with those born in the era where technological advances are evolving by leaps and bounds. The majority of the students we educate have greater access to technological tools, while their trainers are older than the time of the transition in which the experiences with which we pretend to enrich the learning contexts were alien. Mathematics as other sciences, in any form action or any form of use "are the need to be carefully critiqued. This applies to any form of mathematics: mathematics, engineering, math every day, academic mathematics and Ethno mathematics, among many other varieties" [8]. This is, Mathematics, as another science, needs to be adapted to different scenarios, from any of its interpretations, and from its different uses, such is the case of the training of mathematics teachers. The inclusion of this type of environment, involves a greater effort for those who belong to the time, because it implies the recognition of the tool and the identification of possible help by involving them in the educational format that aims to promote. This is, "e-learning has rapidly become a 'must-do' across Higher Education (HE) (...) e-learning is now characterized by virtual learning environments (VLE), usually operating within a managed" [9]. In any case, to talk about involvement with the courses and virtual environment, it is necessary to talk about the motivation to participate in this. In this sense, motivation is something that "should be strengthened in virtual courses, where teachers should strive to maintain it at a high level and develop collaborative activities in an appropriate manner, in addition a relationship between motivation and satisfaction of the course was found" [10]. In particular, the courses with prospective teacher of child pedagogy and mathematics, the participation was greater due to the proportion of use of the virtual tool TelEduc. That is, the courses were programmed so that they made constant deliveries of reports, planning, learning memorials, and prototypes through the virtual platform, which allowed the students to be more motivated to participate. The opposite occurred in the training course of prospective teacher of Physics and Chemistry, where the tool was underutilized when consulting bibliographic material and short reports of the consultations. The more future teachers are involved in the use of these tools, the more products delivered by students that show progress in their learning processes.

The other hand, "the design of several current educational environments increasingly reflects the constructivist pedagogy, by which learners actively construct and interrelate knowledge and ideas" [11]. This is, providing students with rich environments of different elements, such as VLE, allows contact with them to become familiar, and possibly be contemplated by them in future experiences with their own students. On the other hand, it could be demonstrated that the VLE TelEduc was shown as a support to the teaching, in the sense that besides the storage of information of two courses at the same time, the tools could be alternated together with debates and socializations of works, in a way face-to-face with all group participants. Finally, the e-learning model is based on collaborative learning, increase "recognition of the need for education to be closely aligned with team playing, project-based learning and problem-solving approaches has increased the interest in collaborative learning among university and college instructors" [12].

4. Conclusions

Subjecting prospective teachers to different learning environments, allows them to experience other paths in teaching, when they become teachers in exercise. In other words, the lack of access to experiences in VLE, Communities of practice, other physical spaces, possibly perpetuates the teaching model lived by the future teacher. For the trainer of future teachers, it implies the recognition of a need for change in the educational paradigm, as well as an insertion in it, thinking in Accessibility, usability and navigation become essential elements to take into account when carrying out a review of the measurement of e-learning in educational environments. As an aim conclusion, we can highlight the prospective teacher had an investigative stance on their own practice providing a broader understanding of aspects before and during the participation that contributed to their professional perception as prospective teachers. Performing exploratory activities requires a greater effort from the trainers and a different posture than one is accustomed to in the classroom. As was worked out the activities allows context of reflective and investigative participation for the prospective teacher in mathematics, physics, chemical and Early Childhood practices of teaching and learning in the School. Finally, the virtual learning environment can be a pretext for constituting learning communities for teachers and future teachers, who participate to consolidate professional teacher learning.

Acknowledgements

The authors would like to thank the University of Campinas (Brazil) for allowing to explore in appropriate research environments in higher education.

References

[1] Cardona Román D M and SánchezTorres J M 2010 *Revista UIS Ingenierías* **10** 3–52
[2] Clarke L 2013 *Technology, Pedagogy and Education* **22 (1)** 121-131
[3] Reeves T 1994 Evaluating what really matters in computer-based education Eds. Wild M and Kirkpatrick D *Computer Education: New Perspectives* (Perth: Cowan University) pp. 219-246
[4] Bryant S M and Hunton J E 2000 *Issues in Accounting Education* **15(1)** 129–62
[5] Wenger E 1998 *Communities of practice* (Cambridge: Cambridge University Press)
[6] Rinaldi C 1999 Emerging curriculum and social constructivism Eds. Edwards C, Gandini L and Forman G *The hundred languages of the child Reggio Emília's approach to early childhood education* (Santa Barbara: ABC-CLIO, LLC) pp. 113-122
[7] Acevedo-Rincón J P and Fiorentini D 2017 *International Journal of Research in Mathematics Education* **7(2)** 24-44
[8] Flórez Pabón CE and Acevedo Rincón JP 2018 *Journal SYLWAN* **162(9)** 80-89
[9] Larvin M 2009 E-Learning in surgical education and training *ANZ Journal of Surgery* **79** 133–137
[10] Cardona-Roman D 2011 *Propuesta de diseño de metodología para evaluar el impacto del e- learning en el proceso de aprendizaje de estudiantes de educación a distancia* (Bogotá: Universidad Nacional de Colombia)
[11] Rousos M, Johnson A, Leigh J A, Vasilakis A, Barnes C and Moher T 1997 *Computer Graphics* **31(3)** 62-63
[12] Al-Ammary J H 2013 *International Journal of Social, Behavioral, Educational, Economic, Business and Industrial Engineering* **7(2)** 408-415

IOP Conf. Series: Journal of Physics: Conf. Series **1161** (2019) 012024 doi:10.1088/1742-6596/1161/1/012024

The flipped classroom and its application to financial analysis course: Empirical evidence to improve financial culture

D Prada[1,2], A Acevedo[1], J Mantilla[1], W Figueroa[1], A Diaz[1] and G Ardila[1,2]
[1] GIM, Universidad Pontificia Bolivariana, Bucaramanga, Colombia
[2] SED, Universidad Pontificia Bolivariana, Bucaramanga, Colombia

E-mail: duwamg.prada@upb.edu.co

Abstract. Flipped Classroom, Twenty first century teaching technique, oriented to transfer a slice of the learning component to the person within his comfort zone and/or family environment; it is an activity or transference that pretends to stimulate individual tasking of the learner through the use of the technologic element, in such a way that the face to face course time in the classroom is significant to deeply develop the topics object to study. This initiative is relevant to be applied in the teaching process of financial sciences, particularly the financial analysis learning, activity developed by means of reviewing accounting data, which is vital in order to manage and to understand companies, inasmuch as from the holistic understanding of the company itself, it is possible to build survival business strategies; therefore it is necessary that the apprentice deeply comprehends the business complexity and fragility, also that he can use the analysis to take different decisions to improve the business dynamics, with the application of this methodology within the subject of financial analysis, it was possible to demonstrate that it effectively brought visible, positive and general interest results for the professors and other members of the academic year, visualized these in significant improvements in both numerical and of participation and commitment on the part of the student.

1. Introduction

The flipped classroom technique poses its roots in the blending of e-learning (asynchronous learning) and face to face meetings (synchronous) denominated by the education slang as Blended Learning, this technique harmonize greatly the face to face teaching along with technology, it means that its purpose is to alternate some learning tasks with others aided by multimedia components of individual approach and with specific goals and objectives [1]. The value proposal of the technique lies in the concept that each student can appropriate knowledge and review the contents at a pace adjusted to each personal condition, added to this, the learning activities could be held in the preferred place, as many times as the learner wishes, interacting with a studying tool that he dairy uses, for instance YouTube. This learning freedom is extrapolated to the financial analysis subject environment from the business management program where the class is being imparted, finding significant advances and explicit improves in the results, which will be explained precisely in the upcoming paragraphs.

2. Background

There is empirical evidence from numerous authors who have contributed to the object of study. The foregoing focuses in the fact that the student is favored by different didactic strategies that scholars build and propose to improve the students learning. The developed review in holistic way about the flipped classroom subject drops as factual evidence that there is a significant improvement in the learning indexes, in the specific places where the educative model was applied. It is necessary to start from a key concept in the education equation, it needs to suffer a change, in which the professor is no longer seen as the center of the formative process, or the source of fragmented and barely inspiring knowledge, but the principal variable of the education equation is the student. This kind of criteria covers the main idea, which is that this method propose that the actual source of knowledge can be found in the right use of technology. However it is false that the student can truly learn simply by obtaining knowledge through this practice, it is greatly essential the existence of a guide that educates the student so he develops abilities within the competences presented, to accomplish this, the flipped classroom method was described in 2000 for the first time by Baker, who tried to provide the students the learning material for external class and the opportunity to work more in collaboration with teachers and between them during class. According to [2] expose that students must be actively involved in class, since the flipped classroom offers more opportunities to the students to develop a critic and independent thinking and to improve their own learning processes interacting in collaboration with their classmates. Also, teachers offer flexibility and freedom to the students to learn from specially designed activities to improve their skills at overcoming problems.

By other hand, indicate that flipped considers the classroom as a space where the students have previous knowledge of the topic being subject of the class, and that knowledge was given previously, providing an additional plus to all technology we have nowadays to our disposal, but "A way to encourage commitment from the students and ease active learning is by giving them the opportunity to choose what they will do to study and practice [3].

Authors said [4] precise that flipped classroom methodology move some learning processes of learning in environments off class, activity that potentizes spaces in classroom so the teacher can explain concepts that were not understood, allowing a collaborative learning. The previous approach gets its sustentation from [5] when precise that if structured education within the traditional system of teaching, where learning and its interaction is the transmission of knowledge in one way (professor – student), this process gets truncated, weakening the appropriation of thinking skills. In this very line [6] warn that within traditional models, learners can retain little pieces of curricular content, wasting the learning potential, scenario which can be improved by using video, that lead to debate and move the knowledge.

Consider that teachers should use professional judgement in the same measure they offer options, because limited options are more appropriated for most students that possess absolute autonomy [7]. In this very line [8-10] indicate that a practical way to ease the limited election is through selection tables or list selection, in which the teacher must guide the students, teaching to look for the best information restricting the search spectrum to make sure students get the most optimum learning, to have key tools that they can use and they can enjoy such as videos, since a study proves "students generally have positive attitudes towards videos previously used before course class.

In [11], they identified that the students who received classes with video aids as a complement in education, came to the class more apt than when they received readings from textbooks or traditional learning, but unfortunately, in the observations of [12] they found in their analyzes that university students generally do not comply with the reading tasks, which was precisely found in this study.

Finally and as a reference and incentive to this proposal it is necessary to cite the findings in [13], who developed an exercise of inverted class in the course of computational interaction, for which the students developed learning activities with self-taught content aided by the factor technological, obtaining as results an improvement both in the development of the tasks, as in the workshops and of course an improvement in the results of the exams.

With the previous analyzed approaches, a relevant interest in applying flipped classroom methodology to financial analysis course was generated, course chosen principally for two reasons, this is the spine of business dynamics knowledge, and a second reason is that the university where it was held exhorts teachers to develop activities that potentize learning processes using TICs acquired for the professional labor.

3. Methodology

To develop the present study a timeframe of two semesters was consider, July to December (2016) and January to June (2017), for four courses which have 26 students in average per classroom each semester, 106 students in total. For such a study it was implemented a quasi-experimental design with quali-quantitative approach, since what is pretended is to teach the academic community the obtained results from the teachers and students experiences involved in the learning dynamics.

For this study a measuring instrument was built, which was applied to both cohorts of students that were inscribed to the financial analysis course, during the timeframe previously mentioned, the work looked for axioms in standardized scales like the student cohesion, innovation and academic performance. The previous referred studies were analyzed, along global scoring of the instrument, determining a p. valued ($p<0,05$); in the other hand, to review the consistency of the instrument, it was developed a recognition through the verification of alpha coefficient from Cronbach, to then proceed applying it to the target subject of study, task held in means of using e-platforms. Due to a strict methodologic application of the work, it is considered that the general lecture of the exercise is well aligned with quantitative results.

4. Results

The 106 questionnaires addressed by the students from which men: 58%; age: 19.6±4.13 years and women: 42%; age: 20.5±4.11 years. The global evaluation of the flipped classroom exercise and its variables drops important results, said results provide guarantee of the methodology efficiency proposed to improve the learning process.

Regarding the inferential analysis, the non-parametric statistic applied was the Chi-squared test from Pearson, for independent sampling, with the objective to identify and or establish significate statistical differences among the participating students of the technique measure and the traditional model concerning the self-perception scale of initial learning. To measure significate statistical differences, it was considered a confidence level ($1-\alpha$) of 95% and alpha (α) of 5%. However, one of the three elements evaluated in the exercise was to determine the students' perception with reference to the use of the technique and the appropriation of contents using the technological element, the work was addressed by the Likert scale measurement, with four possible answers; The hypothesis proposed for this stage was the identification of the existence of significant differences between the opinions of the two groups studied, information that can be specified in Table 1 and Table 2.

Table 1. Observed and expected frequencies referent to the actualized content.

Qualification	Observed frequencies			Expected frequencies		
	Student G1	Student G2	Total	Student G1	Student G2	Total
1-Strongly disagree	0	2	2	0.9245	1.0755	2
2- Disagree	1	4	5	2.3113	2.6887	5
3-Agree	15	35	50	23.1132	26.8868	50
4- Strongly agree	33	16	49	28.1981	32.8019	61
Total	49	57	106	49	57	106

Table 2. Analysis.

Chi squared	8.1035
Chi tabled	7.8147
Degrees of freedom	3
P-value	0.0439
Valid cases	106

With a level of significance of 5%, it resulted, a test statistic Chi^2=17.1921, that is, the hypothesis that there are significant differences between the opinion of the groups addressed in this work is rejected. It is concluded that there is no difference between opinions; a p-value =0.0006, evidence that the Ha, is rejected. That is to say, the evidence indicates that the perception of the group studied in the second semester of 2016 and the group of the first semester of 2017 has a high correlation in terms of their appreciations regarding this first scenario.

A second stage that we wanted to review with the research work was the perception of the students with reference to the correlation of the technique with their academic performance; by applying the same nonparametric statistical method Chi^2, and a level of significance of 0.05, the results achieved in this exercise (see Table 3 and Table 4), indicate that the calculated Chi^2 is 8.1035 and the Chi^2 Critical is 7.8147, a result that indicates that the hypothesis regarding significant discrepancies between the groups studied is rejected, meaning that they agree with a high degree of significance that they noticed significant changes that led to improve academic performance.

Table 3. Methodology perception and impact on academic performance.

Qualification	Observed frequencies			Expected frequencies		
	Student G1	Student G2	Total	Student G1	Student G2	Total
1-Strongly disagree	0	2	2	0.9245	1.0755	2
2- Disagree	1	3	4	1.8491	2.1509	4
3-Agree	13	26	39	18.0283	20.9717	39
4- Strongly agree	35	26	61	28.1981	32.8019	61
Total	49	57	106	49	57	106

Table 4. Analysis.

Chi squared	8.1035
Chi tabled	7.8147
Degrees of freedom	3
P-value	0.0439
Valid cases	106

Lastly, to measure the academic performance, the students were asked: Do you consider that a change in the traditional methodology can influence your content retention, improving it? With a significance level of 5% (0.05), it drops as result, a statistical proof Chi^2=0.00 i.e. the hypothesis that states that there are significant differences among the students linked to the research process is rejected. It is concluded that there is not any difference among the opinions from students treated with flipped classroom and students oriented with traditional methodology, the previous one obtains its virtue in the result of the statistic p-value=0.000, which evidences that the Ha, is rejected. (See Table 5 and Table 6).

With a significance level of 5% (0.05), it drops as result, a statistical proof Chi^2=0.00 i.e. the hypothesis that states that there are significant differences among the perception of change of traditional methodology from students. It is concluded that there is not any difference among the opinions from students; a p-value =0.000, which evidences that the Ha, is rejected.

Table 5. Change of methodology and content retention.

Qualification	Observed frequencies			Expected frequencies		
	Student G1	Student G2	Total	Student G1	Student G2	Total
1-Strongly disagree	0	1	1	0.4623	0.5377	1
2- Disagree	1	1	2	0.9245	1.0755	2
3-Agree	22	44	66	30.5094	35.4906	66
4- Totally agree	26	11	37	17.1038	19.8962	37
Total	15	91	106	49	57	106

Table 6. Analysis.

Chi squared	13.8897
Chi tabled	7.8147
Degrees of freedom	3
P-Value	0.0030
Valid Cases	106

5. Conclusions

The use of the technological element to improve the appropriation of knowledge, multiplied this with innovative techniques in the learning processes, are a guarantee of a significant improvement in education, this is the case that was evidenced in this exercise where the application of the Flipped Classroom technique to the course of financial analysis for the program of business administration, obviously led to improve not only the notes of the students, but to implement self-taught methodologies by the student, along with a marked discipline in pursuit of your own learning; in this way, the main variable of the educational equation is linked to its formation process, which is the student along with the level of appropriation of knowledge at a particular speed of each individual; The study showed that the technique worked in synergy with the technological element showed positive results in each one of the study groups, and additionally I see that there is no statistically significant evidence of discrepancy between the perceptions of the groups under study.

References

[1] Staker H and Horn M 2012 *Classifying K–12 Blended Learning. InnoSight.* Consulted on: www.innosightinstitute.org
[2] Turan Z and Goktas Y 2016 The Flipped Classroom: instructional efficency and impact of achievement and cognitive load levels *J. e-lear. Know. Soc.* **12(4)** 51
[3] Flores O, Del-Arco I and Silva P 2016 The flipped classroom model at the university: analysis based on professors and students assessment in the educational field *Int. J. Educ. Technol. High. Educ.* **13** 21
[4] Prieto A, Prieto B and Del pino B 2016 Una experiencia Flipped Classroom *Actas de las XXII Jenui* Almería (Granda: Universidad de Granada) pp. 237-244
[5] Sánchez R 2017 *Aula invertida, metodología del siglo XXI.* (Islas Baleares: Universitat de les Illes Balears)
[6] Anderson L and Krathwolhl D 2001 *A Taxonomy for Learning, Teaching and Assessing: A Revision of Bloom's Taxonomy of Educational Objetives* (New York: Logman)
[7] Sams A and Aglio J 2017 Three ways the Flipped Classroom leads to better subject mastery *Edu. Dig.* **82(5)** 52-54
[8] Long T, Logan J and Waugh M 2016 Students' Perceptions of the Value of Using Videos as a Pre-Class Learning Experience in the Flipped Classroom *TechTrends* **60(3)** 245-252
[9] Strayer J 2012 How learning in an inverted classroom influences cooperation, innovation and task orientation *Lear. Env. Res.* **15(2)** 171-193
[10] Tourón J and Santiago R 2015 El modelo Flipped Learning y el desarrollo del talento en la escuela *Revista de Educación* **368** 196-231
[11] Janet L, DeGrazia J, Falconer L, Garret N y Will M 2012 Incorporating screencasts into chemical engineering courses. P*roceedings of the ASEE Annual Conference & Exposition*
[12] Sappington J, Kinsey K and Munsayac K 2002 Two studies of reading compliance among college students *Teach. Psyc.* **29(4)** 272–274
[13] Day J and Foley J 2006 Evaluating a web lecture intervention in a human–computer interaction course *IEEE Trans. on Edu.* **49(4)** 420–431

IOP Conf. Series: Journal of Physics: Conf. Series **1161** (2019) 012025 doi:10.1088/1742-6596/1161/1/012025

A model of didactic guidelines to train designing and implementation of biodigesters

J Quiroga[1], **K Tschiersch**[1] **and O Bohórquez**[1]
[1] GIEMA, Universidad Industrial de Santander, Bucaramanga, Colombia

E-mail: jabib@uis.edu.co

Abstract. Biodigesters can be a suitable solution to exploit organic waste, fermenting it into biogas and biofertilizer. The presented type of biodigesters is easy to build, to operate, and to maintain even by nonqualified personal. In this document, guidelines for the design, as well as for the installation and for the operation of a small biodigester in rural areas is presented. The most important components and their function in the process of organic digestion are also described. The calculation is simplified to an iterative process to determine the geomembrane dimensions for a given daily organic waste production.

1. Introduction

In rural areas, waste management is not a very common task. The inadequate residues disposal can cause sanitary problems that burden the environment and people´s health [1]. Nowadays, biodigesters are gaining the attention of the researching community due to their low cost and effectivity in transforming the organic residues into a valuable product. In general, a biodigester converts organic material in biogas [2], which can be used, e.g. for cooking or heating applications, and, liquid biofertilizer, which provides suitable nutrients and therefore improves plant growth. The organic material may be human or animal excretes [3], or vegetable waste. The biodigester is a closed, hermetic, and impermeable container, in which the organic material is deposited to ferment (while producing biogas) due to microorganisms [4].

Biogas is a gaseous combustible, formed mainly by methane (CH_4) and carbon dioxide (CO_2). Although, the composition mainly depends on the type of organic residue and the developed microorganisms in the digestion stage, in Table 1 the average composition in percentage by volume of biogas is presented.

Table 1. Percentage in volume of main components of biogas [5].

Components	% in volume
Methane (CH4)	55 − 65
Carbon dioxide (CO2)	34 − 45
Nitrogen (N2)	0 − 3
Hydrogen (H2)	0 − 1
Hydrogen sulfide (S2H)	0 − 1

A biofertilizer is a liquid material, rich in nitrogen, which contains considerable amounts of potassium, phosphorus, and other elements, essential for plant development. In addition to the low-cost, another advantage of using natural fertilizers is a better absorption of its nutrients than those of the unprocessed organic matter or even chemical fertilizers by the plants [6].

IOP Conf. Series: Journal of Physics: Conf. Series **1161** (2019) 012025 doi:10.1088/1742-6596/1161/1/012025

Simultaneously to the production of biogas and biofertilizer, the contaminating potential of the excrement is reduced and therefore the groundwater is protected. The whole process achieves an "upgrading" of the biological waste, while making it a product. Also, an economic impact is accomplished.

In this paper, a step by step design procedure of a biodigester is explained, to produce a meaningful learning of thermodynamic and chemical concepts for engineering students.

2. Methods

To decide which system takes the most advantage of organic residues in rural areas, first of all, the flow type of the organic material needs to be studied. Discontinuous-flow means that all the material to be processed is loaded at the beginning. Once the fermentation process has been carried out, the biofertilizer is extracted and the loading is performed again. There is no biogas production in this type.

The semi-continuous flow implies loading and unloading of the material in small lapses (usually once a day). They are the most used biodigesters in rural areas when they are destined for domestic use. The most popular designs are: Chinese type, Hindu type and Taiwan type [7].

The continuous flow biodigesters are generally implemented in industrial scale for the treatment of wastewater. They are installed with heating devices, agitation and power control, and produce large amounts of gas.

Due to the aforementioned, a semi-continuous flow biodigester is the most common installation in rural areas. Therefore, choosing between the following 3 systems is necessary:

- Chinese: fixed dome biodigester.
- Hindu: floating dome biodigester.
- Taiwan: tubular membrane horizontal biodigester.

In this paper, a Taiwan biodigester will be treated, as shown in Figure 1. The tubular membrane is made out of a completely sealed high-density plastic. The lower 75% of the biodigester volume contains the material, which is being fermented and the upper 25% contains the biogas. The microbial fermentation process is completely anaerobic (no air is included in the process after the starting mechanism).

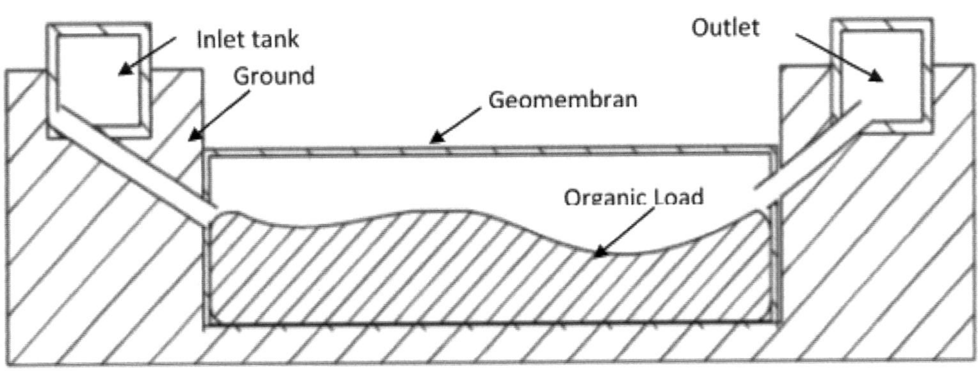

Figure 1. Scheme of a Taiwan biodigester [8].

The most important factors to guarantee a good operation of the process are temperature, PH, and water content in the dilution. Temperature increase accelerates the degradation of the material and the production of biogas. Three temperature levels of anaerobic digestion occur: syphilic, mesophilic and

IOP Conf. Series: Journal of Physics: Conf. Series **1161** (2019) 012025 doi:10.1088/1742-6596/1161/1/012025

thermophilic. But even more important than the temperature level, is maintaining it constant since fluctuations negatively affect the bacterial action.

A neutral medium guarantees the survival and reproduction of methanogenic bacteria. Experience shows that the PH must be maintained between 6 and 8.3 during the entire biogas production process.

Lastly, the dilution percentage of the organic material is an important factor, since bacteria and other microorganisms are unable to function effectively for low water content in the mix. Due to the high-water content of excrements, a proportion between the mixture and water of 1:1 is selected. The dilution of the mix can be adjusted depending on the necessity of establishing a flow to the biodigester.

3. Results (guidelines)

A guideline with the most important steps to follow for realizing a biodigester system will be presented.

3.1. Collection of information

The biodigester is basically formed by a tubular flexible geomembrane, an excavated ditch for geomembrane position, and a safety valve to limit system pressure.

Inside the flexible geomembrane, a displacement of the mixture from the inlet to outlet occurs, while its transformation into fertilizer and biogas take place. Each new part of the diluted mix, that is added to the biodigester, pushes the next part and so on, displacing the mass towards the other end until the completely processed material reaches the biodigester outlet. The remaining volume within the geomembrane will be reserved for storing the produced biogas (see Figure 2). The whole dimension of the process depends on the quality of available organic material, not on the necessity of biogas nor fertilizer. To dimension the system some data is required: the amount of daily excrement, type of excrement, and the average temperature of the location. In Table 2 the percentage of produced biogas depending on material type and days of fermentation is shown.

Table 2. Days of fermentation and percentage of produced biogas depending on material at 30°C [10].

Days of fermentation	10	20	30	40	50	60	70	80	90	100	Generation rate [m^3/kg]
Material	Percentage of the total volume of the produced gas										
Human excrements	40.7	81.5	94.1	98.2	98.7	100					0.478
Pig excrements	46	78.1	93.9	97.5	99.1	100					0.405
Cow excrements	34.4	74.6	86.2	92.7	97.3	100					0.3
Green grass				98.2		100					0.41
Wheat straws	8.8	30.8	53.7	78.3	88.7	93.2	96.7	98.7	100		0.435

3.2. Design of the biodigester

With the provided information, the values of volumetric fluid, retention time, and required liquid volume are calculated [7]. The volumetric flow per day $\dot{V}$ is determined by Equation (1), where Q_{excre} is the flow rate of organic material in a day, obtained from collected information in the location, and Q_{water} the flow rate of the amount of water (proportion about 1:1 is recommended).

$$\dot{V} = Q_{excre} + Q_{water} \tag{1}$$

The retention time (t_r) is the required time to degrade the organic material and it is directly related to the ambient temperature and the type of used material. A retention time range of a few to 40 days is recommended. Under optimal conditions, around 30°C, retention time should be about 20 days but due to varying temperatures, i.e. decreasing values with respect to the optimal, retention time will increase in a range between 30 to 45 days or longer. Therefore, when a more retention time is assumed means a more conservative design, i.e. a more possibility to complete processing of the waste. The required liquid volume V_{req} results in Equation (2).

$$V_{req} = \dot{V} * t_r \tag{2}$$

Figure 2(b) shows the recommended shape of the perimeter of trench and geomembrane. The transversal perimeter is given by Equation (3). Out of stability matters, the relation of $^x/_h = 0.3$. a is the distance between the walls of the biodigester and is defined in Equation (4), in the same way another geometric variables are shown in Equations (5) and (6).

$$2\pi r = b + 2\sqrt{h^2 + x^2} + \theta r \tag{3}$$

$$a = r\sqrt{2 - 2\cos\theta} \tag{4}$$

$$A_{liq} = \frac{b+a}{2}h \tag{5}$$

$$A_{gas} = \frac{r^2}{2}(\theta - \sin\theta) \tag{6}$$

Where A_{gas} is the gas-filled area and A_{liq} the mix-filled area. The relation of $A_{liq}/A_{gas} = 3$ is applied, as earlier mentioned. The length of the geomembrane is calculated by Equation (8). In order to simplify the calculation at the design stage, distance x is assumed as zero and using the relation in areas between A_{gas} and A_{total} which is being written as

$$\frac{A_{gas}}{A_{total}} = 0.25 = \frac{1}{2\pi}(\theta - \sin\theta) \tag{7}$$

With θ as a unique variable. Now, an iterative process is accomplished in Equation (7) to determine θ which is approximately 132.45°. Thus, replacing θ in Equations (4) and (6) to obtain c and A_{gas}, expressed as a function of r and given by $a \approx 1.83r$ and $A_{gas} \approx 0.78r^2$. Then, knowing that $A_{liq} \approx A_{total} - A_{gas} \approx 2.35r^2$, where A_{liq} is assumed occupying a rectangular space with c and h as width and height respectively, it is also possible to express h in function of r as $h \approx 1.28r = A_{liq}/c$. The estimated biodigester length is calculated by the total volume relationship given as.

$$V_{total} = \frac{V_{liq}}{0.75} = (A_{liq} + A_{gas})L_{geo} \tag{8}$$

Recalling that $V_{liq} = V_{req}$ and replacing the total area in Equation (8), L_{geo} can be written as the Equation (9).

$$L_{geo} = \frac{V_{req}}{0.75(\pi r^2)} \tag{9}$$

In the calculation process, r is assumed to estimate L_{geo}, which is validated using the Equation (10).

$$10 > L_{geo}/2r > 5 \tag{10}$$

Equation (10) is recommended by economic consideration (the smaller the superficial area of the geomembrane, the lower the cost) and operational aspects [7]. In summary, all the information given, is used initially to estimate the unknowns (r, L_{geo}). Next, the dimensioning process of the trench (a,b,x,h) is completed.

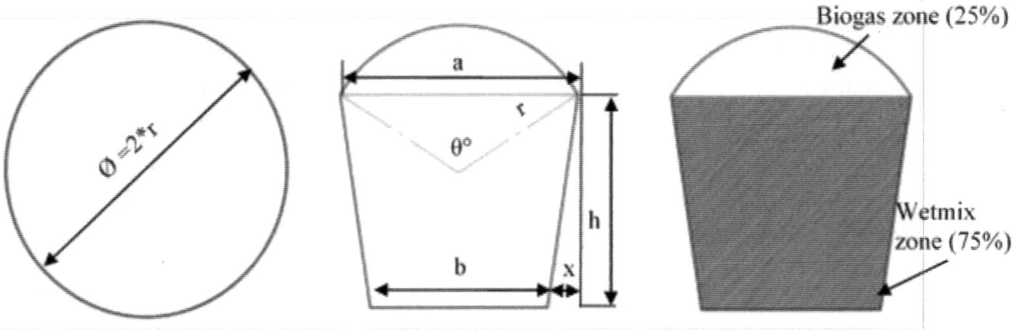

(a) (b) (c)

Figure 2. Transversal profile of the geomembrane, (a) tubular flexible membrane, (b) trench mounted flexible membrane, (c) distribution of areas.

As a calculation example: biodigester load V_{req} =60l, (mixture of organic material and water), assuming a retention time of 50 days and a radius of 0.45m, produce a L_{geo} of 5.09m, Equation (9), which satisfied the Equation (11). Thus, the data for implementation is shown in Table 3:

Table 3. Biodigester construction data (example).

Variable	Expression	Value [m]
L_{geo}	Eqs (9-10)	5.09
r	Assumed	0.45
h	1.28 r	0.652
a	1.83 r	0.822
x	0.3 h	0.196
b	a – 2x	0.431

3.3. Implementation and installation

The system needs two tanks: the first is utilized to mix the organic material with the appropriate amount of water, this tank will be connected by PVC tubes to the loading port flange of the geomembrane; the second tank is used to collect the produced liquid biofertilizer in the geomembrane outlet port. The tank size needs to be at least of the size of the biofertilizer production of one day. The loading tank needs a wire netting to impede blockades. In the following, the security devices, which should be included in an installation, are shown:

3.3.1. Relief valve. this device relieves biogas to the surroundings if the biodigester internal pressure is above of a determined pressure. A low-cost relief valve is proposed based on a plastic T-piece and a pet bottle (0.5 l). The biodigester internal pressure is adjusted by the water column in the bottle, a 0.15 m of water column can be enough to control the internal pressure in the system.

3.3.2. Hydrogen sulphide filter. Hydrogen sulphide is a highly corrosive compound present in the biogas. Therefore, it needs to be withdrawn by a filter installed in the transport line of biogas. For a home-size system, a filter is implemented using a pipe portion of a diameter at least one inches bigger than the size of the transportation line and filling this tube with steel wool. A 0.1 m of length pipe is in most of the cases adequate to remove of the stream this compound.

3.3.3. Carbon dioxide filter. The elimination of carbon dioxide improves biogas quality. For this type of biodigesters, which are operating at very low pressures, filters of receptacles, plastic or glass bottles are recommended. It is filled with an alkaline solution (water + calcium hydroxide): 2g of calcium hydroxide per 1l of water. The filter is normally installed between the biodigester and the hydrogen sulphide filter in a fixed place, hermetically sealed (see Figure 3).

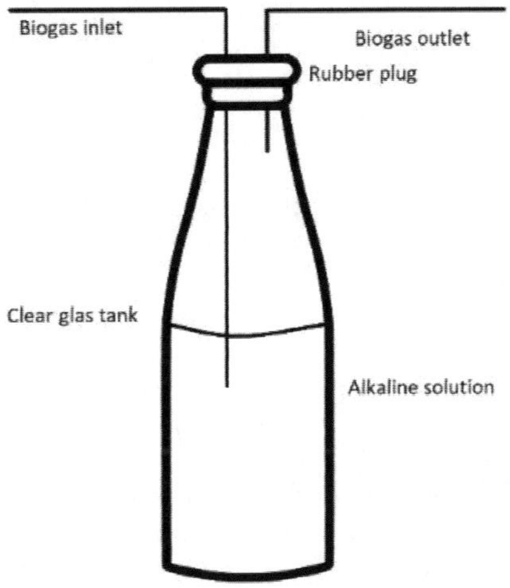

Figure 3. Carbon dioxide filter.

3.3.4. Water trap. Normally, the biogas is a wet gas with a very low vapor pressure, thus at room temperatures, the condensation can be produced. In order to avoid blockages due to condensation, a water trap is required to be installed at the lowest point of the pipeline. The water trap is a water-filled tube, to avoid escape of biogas, as shown in Figure 4.

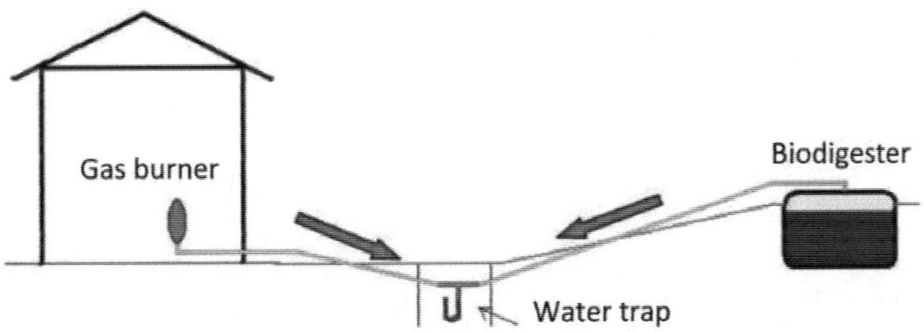

Figure 4. Installation of water trap[13].

3.4. Biodigester start-up

Before the biodigester reaches the steady-state operation, some steps have to be performed. First, the geomembrane needs to get inflated, next the geomembranes feeding, and outlet flanges need to be locked. Later, it will get filled with water, allowing the escape of air through the safety valve, until reaching the hydraulic seal [14]. (Add: From this phase, biogas and biofertilizer can be taken and brought to a laboratory. Inhibitors as shown in **¡Error! No se encuentra el origen de la referencia.** need to be avoided).

3.5. Operation

Finally, to combust the biogas an adjusting process in the burner is required. Thus, the nozzle diameter of the injection valve in the stove is increased step by step until the flame presents similar characteristics to the obtained by LPG or NG by a trial and error process.

4. Conclusions
A biodigester exploits organic waste and produces biogas and biofertilizer while reducing the possible contamination of groundwater due to excrements filtration. In this document, basic equations and general information is provided, thus a functional biodigester can be designed, installed, tested and put in operation by nonqualified personal.

References
[1] Hernández Embate A J. 2013 *El potencial de los biodigestores como técnica sostenible para la producción de biogás en la comunidad indígena Nuevo San Juan Parangaricutiro, Michoacán* (Michoacan: Universidad Autonoma del Estado de México)
[2] Severiche C A, Acevedo R L2013 Biogás a partir de residuos orgánicos y su apuesta como combustibles de segunda generación *Ingenium - Revista de la Faultad Ingenieria* **28** 6–15
[3] Reuss N, Busso M 2000 *Diseño de un biodigestor para obtener metano utilizando excremento de vacas y cerdos en una escuela agrotécnica de la UNNE* Consulted on: http://exa.unne.edu.ar/investigacion/energia_solar/PUBLICACIONES/Biogas.pdf
[4] Filippin C Follari J and Vigil J 1979 *Diseño de un biodigestor para obtener gas metano y fertilizantes a partir de la fermentación de excrementos de vacas lecheras* en la facultad de agronomia de la Universidad Nacional de la Pampa Consulted on: http://www.produccion-animal.com.ar/Biodigestores/02-Biodigestor.pdf
[5] Stamatelatou K, Antonopoulou G, lyberatos G 2010 12. Production of biogas via anaerobic digestion EDS. Luque R, Camelo J and Clark J *Handbook of Biofuels Production* (Sawston: Woodhead Publishing)
[6] Canepa J R L, Olivier J A S, Moguel K C C, Snachez L J and Olivares J F 2011 Diseño, construcción y operación de un biodigestor anaerobio tipo cúpula a escala real para la obtención de biogás *Hacia la sustentabilidad: Los residuos sólidos como fuente energía y materia prima.* **2(21)** 612–616
[7] Walsh J, Ross C, Smith M, Harper S, Wilkins A 1988 *Handbook on biogas utilization* (Georgia: Georgia Tech Research Institute)
[8] Guardado C 2007 *Diseño y construcción de plantas de biogás sencillas* (La habana: Cubasolar)
[9] Botero R, Preston T R 1978 *Biodigestor de bajo costo para la produccion de combustible y fertilizante a partir de excretas* (Cali: Centro Internacional de Agricultura Tropical)
[10] Guevara Vera A 1996 *Fundamentos básicos para el diseño de biodigestores anaeróbicos rurales. Produccion de gas y saneamiento de efluentes* (Lima: CEPIS)

IOP Conf. Series: Journal of Physics: Conf. Series **1161** (2019) 012026 doi:10.1088/1742-6596/1161/1/012026

Measuring the radius of the earth. Integration between didactic transposition and information and communication technologies

F Durán-Flórez[1], L C Caicedo[1], D A Prada[2,4], A Acevedo[3] and J Gómez[5]
[1] Grupo de investigación en Tecnología del Plasma, Universidad Pontificia Bolivariana, Bucaramanga, Colombia
[2] Grupo de Investigación en Materiales· Universidad Pontificia Bolivariana, Bucaramanga, Colombia
[3] Grupo de Investigación en Administración, Universidad Pontificia Bolivariana, Bucaramanga, Colombia
[4] Grupo de Investigación Saber Educación y Docencia, Universidad Pontificia Bolivariana, Bucaramanga, Colombia
[5] Centro de Gestión de Mercados, Logística y Tecnología de la Información, Servicio Nacional de Aprendizaje (SENA), Bogotá, Colombia

E-mail: fernando.duran@upb.edu.co, duwamg.prada@upb.edu.co

Abstract. The activities that foster a teaching mediation based on continuous reflection, involving didactic transposition and information and communication technologies, have generated a space for construction on the part of the student, which transcends outside the classroom and through the use of said technological tools is possible a meaningful and lasting learning. Said didactic transpositions, technological tools and integrating activities not only generate knowledge of value but also may involve the development of another type of competence and intelligences that are not the objective of the course, but that help in the integral formation of the student.

1. Introduction

Teacher mediation in the basic university cycle subjects is continuously strengthened with the different strategies proposed by teachers under the reflection of their pedagogical work. These strategies become significant when the environment is involved in the planned activities, which are useful to achieve the learning objective in the student. The information and communication technologies (ICT) have become a tool of great importance, not only for the support of presence or the development of technological devices, but because it allows the student to demonstrate what was analyzed in class through simulation, as well as everything what the teacher wants the student to understand so that he is competent both in the theoretical part, as the procedural part and in the argumentation of the observed. This type of activities in which the student is expected to learn a knowledge mediated by a didactic site and not through the use of analogies generates significant learning in the student, this process is known as didactic transposition [1]. The classes that integrate the didactic transposition with the use of technological tools frame a new idea of generating in the student a learning or knowledge transposed under a didactic site mediated by the ICT. An integrated class gives meaning and practical significance to computer science, as it seeks the integration of knowledge in the education of the learner [2], without the use of them excludes the valuable contribution that implies to exchange in the classroom, not only knowledge but the same experiences and ideas [3].

The activities of each subject show the relevance of the didactic transposition based on the specific topics declared in the curriculum. Performing activities in which other subjects are articulated, which in some way are related to the thematic focus that is addressed in class, is a commitment to learning that is not only significant but lasting over time, because the mainstreaming based on the didactic transposition mediated with the ICT, demands both the student and the teacher a more detailed interaction with the learning process. The teaching and learning processes have been evolving as an activity whose objective should be formed in an integral way to people with aspirations to play a professional role within the society based on meaningful learning [4]. The activity developed is supported by the directed research seen as a bet to rethink teaching-learning processes [5], because to achieve conceptual, procedural and attitudinal changes in students, it is necessary to place them in a context of action that emulates the scientific activity always guided by the teacher [6]. The directed research defines as a goal to promote in the students changes in their systems or conceptual, procedural and attitudinal structures, where the curricular axis of the approach is the solution of problems generated in the knowledge of the discipline [7]. From this perspective, the incessant search for theoretical and methodological models that make it possible to understand implicit processes in quality education that call young people to participate actively in society and to engage professionally with it is relevant [8].

For the application of this activity with the students, it was determined to use the theory based on problematic situations which arise in those cases in which the lack of correspondence between the knowledge systems that the students already have and the requirements that it poses to the time to solve new tasks exposed by the teacher [9].

An activity that relates thematic axes of the subject of Mechanics or Physics I, with subjects such as Geometry, Trigonometry and Differential Calculus was applied to students of the Universidad Pontificia Bolivariana in Bucaramanga in 2017. The activity was based on radio measurement of the earth through a methodology used by Eratosthenes, 275 B. C., who taking measurements of the sun's elevation in two different places, provided strong evidence that confirmed the spherical shape of our planet Earth and calculated its diameter with 99% accuracy [10], Eratosthenes supposed that the rays of the Sun were parallel and that when this was on Alexandria, the rays formed an angle of 7.5° with respect to the horizontal in Siena city [11,12]. Being able to demonstrate the discovery of Eratosthenes through the use of technological tools, generates in the students diverse questions that will allow him to acquire a commitment with his own learning.

The basic study necessary to develop the activity proposed for the class, requires only one trigonometry course, Trigonometry is one of the courses of the area of mathematics of wider applicability in physics, engineering, chemistry, aeronautics, among others [13], the trigonometric functions because of their journal behavior has been used for modeling seismic waves, sounds, in medicine they have used trigonometry for several studies that involve the implant inclined to patients [14].

A basic principle of teaching mathematical methods is to start using examples, so students can easily follow the technique [15]. With the generation of these spaces, teaching mediation can become the canal's construction of new theories, the use of known laws helps to discover others using basic methods such as algebraic substitution [16].

2. Methodology

In the subject Mechanics of the Physics Area, attached to the Department of Basic Sciences of the Pontifical Bolivarian University, the physics and mathematics teachers proposed an activity framed in the pedagogical strategy of problem-based learning, in which the students must measure the radius of the Earth. The students form working groups, of maximum 4 people, in order to improve the argumentation according to the calculation of the radius of the Earth through the use of a technological tool, among which are Google Earth, global positioning systems (GPS), cellular, among others.

Based on the idea of Eratosthenes, students choose two cities of their choice and calculate the distance between them with the help of any ICT tool, in this way it is possible to minimize the percentage of measurement error of the radius of the Earth. Basically, the process used by the groups was choice of the object to project the shadow, choice of cities to carry out the simultaneous experiment, measurement of shadows and angles present in the experiment and finally, measuring the distance between cities.

Regarding the choice of the object used by the members of each group, eight of the nine groups chose an object to project a wooden stick, whose measurements were between 1m and 1.8m, and a single group chose to make a scale building which had a height of 50cm. Similarly, students in each group chose cities based on the fact that the experiment must be carried out simultaneously to avoid introducing a new error caused by the time factor, which used cities where students lived their relatives or friends. Once the cities were established, an hour was chosen for each group to carry out the experiment, in which known people interact with the members of each group at the same time. To guarantee the simultaneity of the experiment, cellular communication was established between people located in different cities, we proceeded to measure the height of the object located to the sun and the shadow projected by it, through these data we can know the angle of incidence of the sun's rays. To measure the distance between the cities, some groups used the GPS of their cell phones to find the position in coordinates of the cities, with which it was possible to establish the distance between the cities with the help of Google Earth. Other groups made this measurement directly in Google Earth, for this positioned the mouse cursor in each of the cities with which the Google Earth platform shows the distance between them. Figure 1 is showing some of these distances,

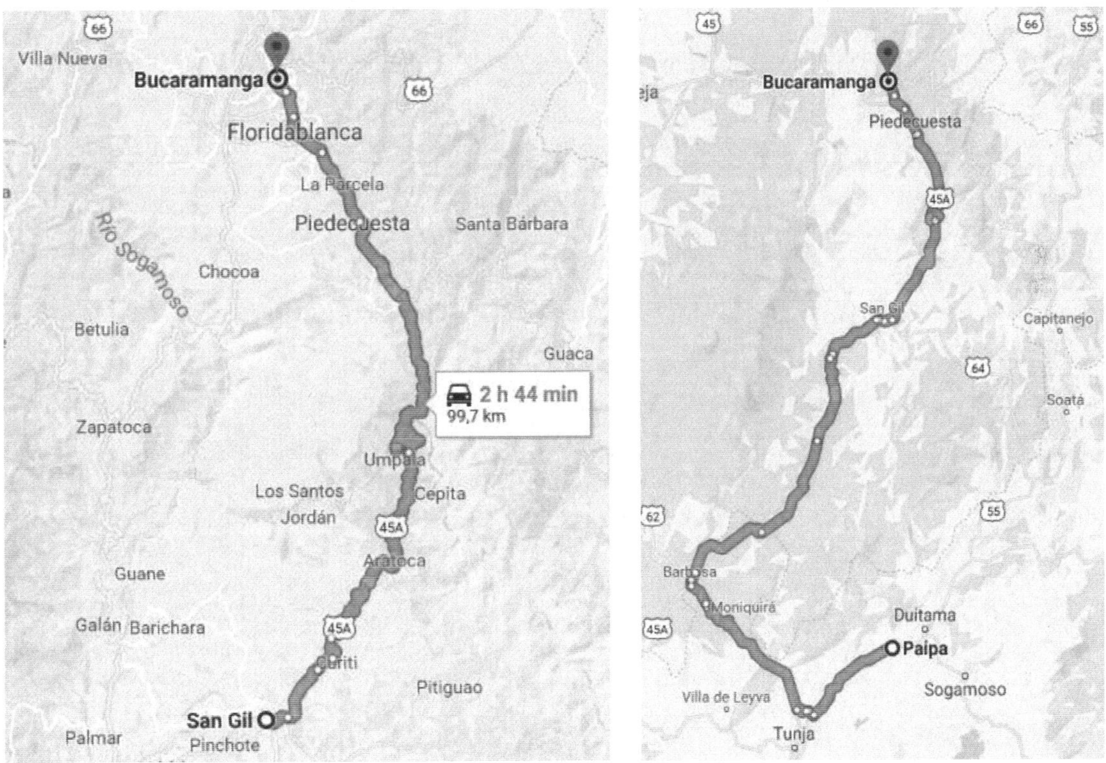

(a) (b)

Figure 1. (a) Image of the department of Santander, Bucaramanga-San Gil. (b) Image of the department of Santander, Bucaramanga, Paipa. Taken from Google Earth.

IOP Conf. Series: Journal of Physics: Conf. Series **1161** (2019) 012026 doi:10.1088/1742-6596/1161/1/012026

3. Conclusion

The students who chose the cities of Bucaramanga, San Gil, evidenced a distance between the cities of 62.11km with an angle of $0^{\circ}35'45''$, perimeter of 37121.79km to find a radius of 5908.11km for a percentage error of 7.38%. The students who chose the cities of Bucaramanga to Paipa, showed a distance between the cities of 147.69km with an angle of $1^{\circ}22'45''$, perimeter of 38549.03km to find a radius of 6135.27km for a percentage error of 3.8%. With the integrating activity it was possible to show that the students' learning was based on the didactic transposition that allowed the measurement of the Earth's radius, however, as more people were involved in the mediation, through ICTs, other types of competencies were observed. Students developed, in which it can be asserted that meaningful learning involves more than the tools and techniques of learning, it also involves social competence.

References

[1] Cantoral R. and Farfán R. 2004 *Desarrollo conceptual del cálculo* (Bucaramanga: Thomson Learning)
[2] Andrade H and Gómez L 1997 *Tecnología informática en la escuela* (Bucaramanga: Universidad Industrial de Santander)
[3] Jaimes M 2009 *La universidad una institución que aprende Revista Docencia Universitaria* **10** 111-121
[4] Salcedo G 2013 Aplicación de los principios éticos de la investigación científica en los procesos de enseñanza y aprendizaje en la educación superior *Revista Docencia Universitaria* **14** 17-28
[5] Velazco F 2012 *Investigación dirigida como modelo en la enseñanza y el aprendizaje de las ciencias naturales. Caso estudiantes del sexto grado de la institución educativa La laguna sede E el regadero* (Bucaramanga: Universidad Industrial de Santander)
[6] Pozo J and Gómez M 2006 *Aprender y enseñar ciencia* (Madrid: Morata)
[7] Quijano M 2012 Enseñanza de la ciencia: retos y propósitos de formación científica *Revista Docencia Universitaria* **13** 17-34
[8] Ravanal E and Camacho J 2012 La actuación pedagógica en docentes universitarios de ciencias: análisis desde el estudio de clases *Revista Docencia Universitaria* **13** 105-136
[9] Majmutov M 1983 *La enseñanza problémica* (La Habana: Pueblo y Educación)
[10] Devlin K 2002 *El lenguaje de las matemáticas* (España, Robinbook)
[11] Clemens S et al 1998 *Geometría* (México: Pearson)
[12] Alsina C 2008 El club de la hipotenusa (Barcelona: Ariel)
[13] Goel S and Elstak I 2015 *Georgia Journal of Science* **73** 36-146
[14] Farideh G and Sadighpour L 2012 Journal of Oral Implantology **38** 289-290
[15] Sher L and Sher D 2007 *Mathemathics and Computer Education* **41** 168-171
[16] Skurnick R and Mohammad J 2006 *Mathemathics and Computer Education* **40** 70-73

IOP Conf. Series: Journal of Physics: Conf. Series **1161** (2019) 012027 doi:10.1088/1742-6596/1161/1/012027

Coil pump design as an object of meaningful learning

J Quiroga[1], K Tschiersch[1] and O Bohórquez[1]
[1] Escuela de Ingeniería Mecánica, Universidad Industrial de Santander, Bucaramanga, Colombia

E-mail: jabib@uis.edu.co

Abstract. Coil pumps can be a suitable solution to solve the pumping problem in isolate communities. This type of pump is easy to build, cheap and the maintenance can be performed for nonqualified personal. In addition, the pumping energy is provided by the river, so no fuel or electricity is required to the pumping process. The previous advantages constitute an attractive subject mainly for students whose water needs are no completely resolved as rural students. The design of this type of pumps demands mainly the application of concepts of fluid mechanics and energy. In this research is proposed a method to teach the design and implementation of a coil pump system in a simple way by a step by step methodology to impel the meaningful learning of the involved concepts in the design stage.

1. Introduction

The lack or the high-cost of conventional energy sources i.e. electricity or fossil fuels in rural areas affect the water supply and consequently its development and compromise the rural survival. Among the possible solutions, non-conventional pumping systems, as the coil pump, is gaining the attention for its low cost and easy implementation. This type of pump works transforming the kinetic energy of the stream (i.e. a river) in potential energy expressed by water high pressure in the pump outlet without the consumption of electricity or fossil fuels.

The coil pump is an easy building system, which is based on evolved prototypes of the Arquimedes pump. The first version of coil pump was proposed by Andrew Wirtz (1749) [1]. Later, in 1979, in the Blair Research Lab in Zimbabwe was designed a coil pump, called a hydrostatic or manometric pump [2]. Next, researches tackled the problem from an analytical perspective. In [3] is proposed a coil pump analytical model to be used in the design and building stages. Based on this a coil pump is tested to determine efficiency as a function of rotational speed, immersion relation and the pumping performance with different hose layers [4]. On the other hand, the determination of the relation between the number of coil layers and the outlet pressure and with the flow rate is investigated in [5] concluding that a more layers of hose more pressure without any effect in the water flow rate. In other research [6] is determined a growth in the outlet water flow rate as a result of two factors: first, an increase of the rotational speed and, second, a rising the submergence ratio, in both cases it is produced the same outlet pressure.

In this work, a step by step design procedure of a coil pump is explained to produce a meaningful learning of thermodynamics and fluid mechanics concepts in engineering students mainly belonging of rural areas, concepts which are involved in the operation of this type of green pump.

IOP Conf. Series: Journal of Physics: Conf. Series **1161** (2019) 012027 doi:10.1088/1742-6596/1161/1/012027

2. Theoretical framework

The coil pump, as shown in Figure 1, is a positive displacement pump which uses its rotation to produce an incremental pressure differential to pump liquids. This pump is composed of a flexible pipe usually a hose wrapped around a cylindrical structure submerged in the fluid to be pumped. The energy supply source to operate this pump is usually obtained from the kinetic energy of the stream, which contains the liquid to be pumped, transformed via a turbine coupled to the main frame of the coil pump.

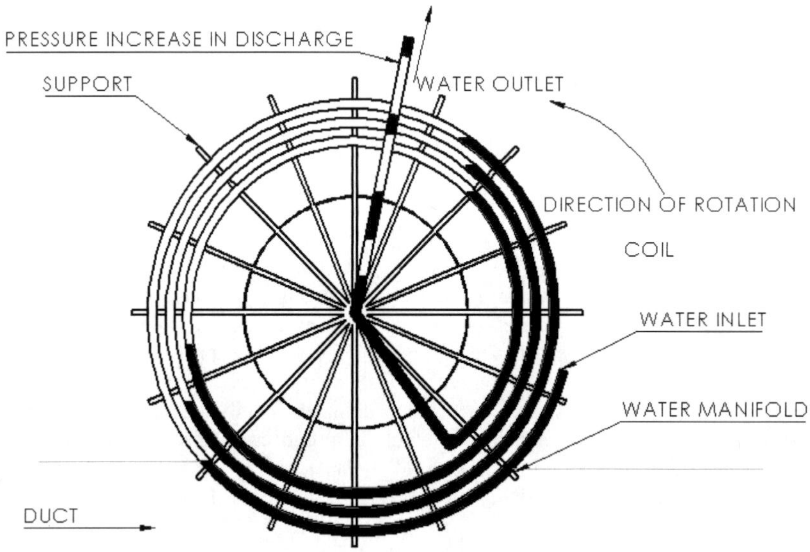

Figure 1. Schematic representation of a coil pump [7].

In operation, coil pumps form a series of alternating water and air packets which constitute a manometers sequence with pressure differentials. The pump outlet high-pressure is the result of the summation of these differentials. Therefore, in coil pumps, the number of hose layers, number of water columns, determines the net discharge head (see Figure 2).

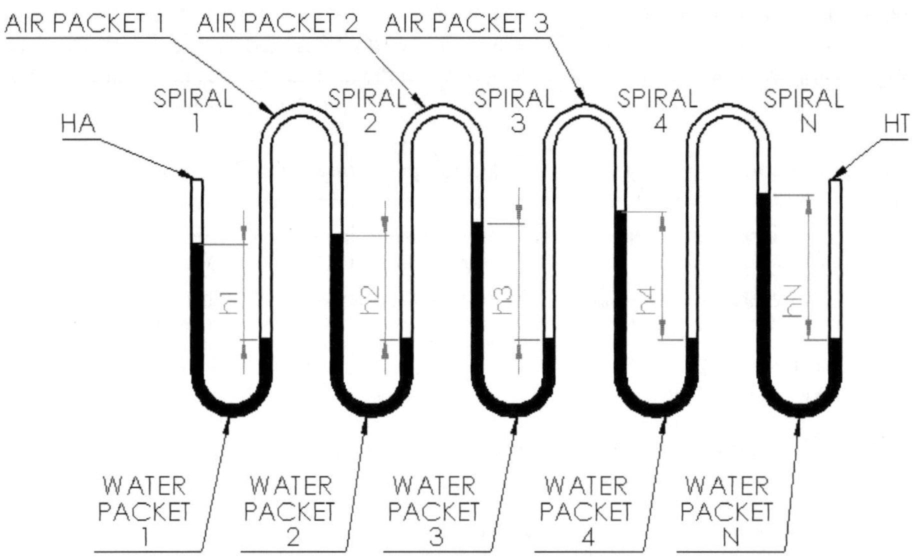

Figure 2. Plugs of water and air inside of hose [8].

The coil pump analytical model to determine the flow rate and the net discharge head was proposed by [9] as presented by Equations (1) and (2).

$$Q_P = N_S \,.\, \pi r^2 .\, L_{W.1} \tag{1}$$

$$H_P = H_T - H_A = \sum_i^N h_i \tag{2}$$

Where Q_P is the outlet flow rate of liquid, r is radius of the flexible pipe, Ns is the cylinder rotational velocity in cycles per minute, and $L_{W.1}$ is the length of the first packet at the pump inlet. Thus, the pump average volume flow rate is defined by the flexible pipe cross-section, submergence ratio defined by the first packet and the pump velocity of rotation.

In Equation (2), H_P is the discharge head, H_T is the absolute net discharge head, H_A is the atmospheric pressure and h_i is the produced pressure differential for each water packet, for the N produced packets inside of the flexible pipe. Air packets lengths are the result of the path of the pump inlet rotating above the water level. The presence of consecutive water and air packets produce an intermittent liquid deliver. These air packets are subject to contraction and stretching due to the different pressures along the flexible pipe around the cylinder from inlet to outlet (see Figure 3).

The initial volume of the air packets, specifically for the first coil depends on the drum geometry, the hose internal diameter and the pump submerged percentage. Because the air packet is limited by the water packets next to it, different air pressures are developed. The air pressure can be studied using a polytrophic relation i.e. $PV^{1,15}$=constant, where P as absolute pressure in the air packet and V is the packet volume.

As the air packets are moving from the inlet to the outlet of the pump, the developed high pressure reduces the air packet volume increasing the water packet volume. This volume change can be calculated using the polytrophic relation mentioned below as stated in Equation (3). When the water packet reaches the crown of the coil a spillback of the water packet can be occurred eliminating the air packet between them. Disappeared air packets involve loss of produced pressure differentials in the coil causing an outlet pressure reduction.

$$H_A L_A^{1.15} = H_n L_{A,n}^{1.15} \tag{3}$$

Where H_n is the absolute pressure head in the n coil, $L_{A,n}$ is the air length packet in the coil n, H_A and L_A are the head and length of the occupied space by the air in the pump inlet. Therefore, the air packet length change between different coils can be determined by Equation (4).

$$L_A - L_{A,n} = L_A\left(1 - \left(\frac{H_A}{H_n}\right)^{0.87}\right) \tag{4}$$

The established flow pattern in this type of pump is characterized by a sequence of air and water packages produced during the operation of the pump. This discontinuous water flow usually unconcern if the facility is for water storage as is the case in most rural communities. In order to properly design a coil pump, a sequence of calculations needed to meet the given requirements. The most important analytical relations are stated as follows:

Equation (5) provides the coil radius in terms of L_{AA}, and L_{BB}, which are related to the length between the packets in the hose per revolution.

$$L_{BB} - L_{AA} = 2\pi R \tag{5}$$

The pump outlet flow is a function of the drum angular velocity, the hose cross-section A_T and the submergence ratio as shown in Equation (6).

$$Q_D = L_{W1} * A_T * rpm \tag{6}$$

Finally, the required torque M under the established pressure conditions is calculated using Equation (7) as follows.

$$M = (H_D - H_A) * A_T * R \tag{7}$$

Where H_A is the atmospheric pressure, H_D is the total pump outlet pressure, and R, is the drum radius.

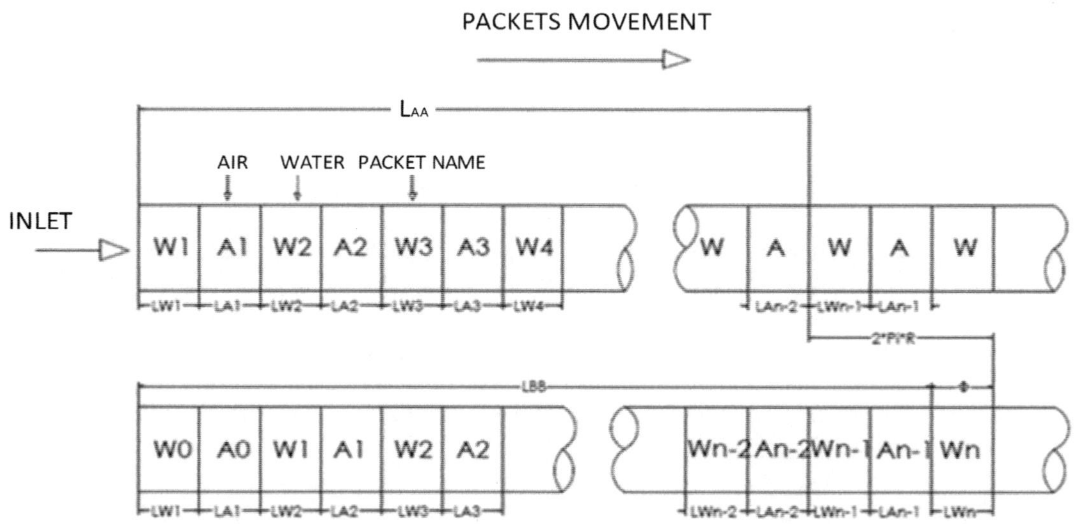

Figure 3. Distribution length of plugs in hose [8].

3. Implementation of a pump coil in a rural area in Colombia

Based on different lab-scale prototypes developed at Universidad Industrial de Santander in the Mechanical Engineering School, a pump coil is proposed, implemented and tested in a rural community in Santander, Colombia. The pump is composed of a metallic frame to hold all the pump elements including the floats[10]. This frame has a metallic bumper impact absorber in the front to attenuate the impact energy of any material carried by the stream. In addition, the coil pump is also made up of the following parts:

A flexible pipe coil: This part is responsible for generating the water and air packets required to pump the liquid. This coil can be built using irrigation pipe which is an economic option easily available in rural areas.

Turbine blades: This component transforms the stream kinetic energy in rotational mechanical energy. These blades can be constructed using transversal cuts of PVC pipes which are suitable, cheap, common and corrosion resistance.

Rotating joint pipe fitting: This part provides a tightness couple between the stationary discharge hose and the rotating coil. Next in the Figure 4, a coil pump schematic representation and an actual picture of the implemented pump is shown, in explode view.

IOP Conf. Series: Journal of Physics: Conf. Series **1161** (2019) 012027 doi:10.1088/1742-6596/1161/1/012027

(a) (b)

Figure 4. (a) Schematic design of installation, (b) Implementation of a coil pump [7].

4. Conclusions

The coil pump design provides a meaningful learning model, due to discuss an essential problem, which is the search for alternative sources of energy to obtain basics conditions like water availability. The coil pumps design includes concepts on disciplines such as thermodynamics and fluid mechanics, which are important in the engineer formation.

On the other hand, a coil pump were designed and implemented to supply water to an isolated rural community in Colombia by undergraduate mechanical engineering students as a part of their thesis. This prototype was used to validate the design process and to satisfy the water needs in this community. The pump was installed in a mighty river that acts as an energy source in such a way that meets the required energy to pump water from the level of the river to a storage tank located 70 m uphill to the stream.

References

[1] Ewbank T 1842 *A Descriptive and historical account of hydraulic and other machines for raising water, Ancient and modern: Including the progessive development of the steam engine* (London: Tilt and Bogue)

[2] Morgan P, McPherson H 1985 The Blair Research Laboratory's contributions to the Water Decade *Waterlines* **4(1)** 9

[3] Kassab S Z, Naby A A A, Sayed E, Basier I A 2005 Coil pump performance under variable operating conditions *Proc. Ninth International Water Technology Conference* (Sharm El-Sheikh: IWTC9)

[4] Naegel, Real, Mazaredo 1991 Designing a spiral pump for irrigation *Waterlines* **10(2)** 30

[5] Patil N, Gaikwad S, Navale R A, Sonawane D S 2013 Design, manufacturing and performance analysis of spiral coil pump *Appl Mech Mater* **446** 549

[6] Hilton D J 1986 The inclined coil pump *Waterlines* **5(1)** 12

[7] Camacho Gomez O 2010 *Sistema de bombeo con energía alternativa para la vereda Naranjal bajo municipo de Socorro* (Bucaramanga: Universida Industrial de Santander)

[8] Pickford J, Mortimer G H, Annable R J 1980 The stream powered rotating coil pump *Water Services* **84** 706

[9] Mortimer G H and Annable R 1984 The coil pump. Theory and practice *Journal of Hydraulic Research* **22(1)** 9

[10] Galvin S 1988 *The flow-media-containing coil pump* (Loughborough: Loughborough University)